Dynamic Optimization and Mathematical Economics

MATHEMATICAL CONCEPTS AND METHODS IN SCIENCE AND ENGINEERING

Series Editor: **Angelo Miele**
Mechanical Engineering and Mathematical Sciences, Rice University

1 **INTRODUCTION TO VECTORS AND TENSORS** • Volume 1: Linear and Multilinear Algebra • *Ray M. Bowen and C.-C. Wang*

2 **INTRODUCTION TO VECTORS AND TENSORS** • Volume 2: Vector and Tensor Analysis • *Ray M. Bowen and C.-C. Wang*

3 **MULTICRITERIA DECISION MAKING AND DIFFERENTIAL GAMES** • *Edited by George Leitmann*

4 **ANALYTICAL DYNAMICS OF DISCRETE SYSTEMS** • *Reinhardt M. Rosenberg*

5 **TOPOLOGY AND MAPS** • *Taqdir Husain*

6 **REAL AND FUNCTIONAL ANALYSIS** • *A. Mukherjea and K. Pothoven*

7 **PRINCIPLES OF OPTIMAL CONTROL THEORY** • *R. V. Gamkrelidze*

8 **INTRODUCTION TO THE LAPLACE TRANSFORM** • *Peter K. F. Kuhfittig*

9 **MATHEMATICAL LOGIC:** An Introduction to Model Theory • *A. H. Lightstone*

11 **INTEGRAL TRANSFORMS IN SCIENCE AND ENGINEERING** • *Kurt Bernardo Wolf*

12 **APPLIED MATHEMATICS:** An Intellectual Orientation • *Francis J. Murray*

14 **PRINCIPLES AND PROCEDURES OF NUMERICAL ANALYSIS** • *Ferenc Szidarovszky and Sidney Yakowitz*

16 **MATHEMATICAL PRINCIPLES IN MECHANICS AND ELECTROMAGNETISM,** Part A: Analytical and Continuum Mechanics • *C.-C. Wang*

17 **MATHEMATICAL PRINCIPLES IN MECHANICS AND ELECTROMAGNETISM,** Part B: Electromagnetism and Gravitation • *C.-C. Wang*

18 **SOLUTION METHODS FOR INTEGRAL EQUATIONS:** Theory and Applications • *Edited by Michael A. Golberg*

19 **DYNAMIC OPTIMIZATION AND MATHEMATICAL ECONOMICS** • *Edited by Pan-Tai Liu*

20 **DYNAMICAL SYSTEMS AND EVOLUTION EQUATIONS:** Theory and Applications • *J. A. Walker*

21 **ADVANCES IN GEOMETRIC PROGRAMMING** • *Edited by Mordecai Avriel*

A Continuation Order Plan is available for this series. A continuation order will bring delivery of each new volume immediately upon publication. Volumes are billed only upon actual shipment. For further information please contact the publisher.

Dynamic Optimization and Mathematical Economics

Edited by
Pan-Tai Liu
University of Rhode Island
Kingston, Rhode Island

PLENUM PRESS · NEW YORK AND LONDON

Library of Congress Cataloging in Publication Data

Main entry under title:

Dynamic optimization and mathematical economics.

(Mathematical concepts and methods in science and engineering; v. 19)
Includes index.
1. Economics, Mathematical–Addresses, essays, lectures. 2. Control theory–Addresses, essays, lectures. 3. Natural resources–Mathematical models–Addresses, essays, lectures. 4. Statistics and dynamics (Social sciences)–Addresses, essays, lectures. I. Liu, Pan-Tai.
HB135.D95 330'.01'51 79-9088
ISBN 978-1-4684-3574-0 ISBN 978-1-4684-3572-6 (eBook)
DOI 10.1007/978-1-4684-3572-6

Softcover reprint of the hardcover 1st edition 1980
A Division of Plenum Publishing Corporation
227 West 17th Street, New York, N.Y. 10011

Contributors

Masanao Aoki, Department of System Science, University of California, Los Angeles, California

Kenneth J. Arrow, Department of Economics, Harvard University, Cambridge, Massachusetts

Tamer Başar, Applied Mathematics Division, Marmara Research Institute, Gebze, Kocaeli, Turkey

Dagobert L. Brito, Department of Economics, Tulane University, New Orleans, Louisiana

William A. Brock, Department of Economics, University of Chicago, Chicago, Illinois, and Department of Economics, University of Wisconsin, Madison, Wisconsin

Sheldon S. L. Chang, Department of Electrical Sciences, State University of New York, Stony Brook, New York

Colin W. Clark, Department of Mathematics, University of British Columbia, Vancouver, British Columbia, Canada

Michael D. Intriligator, Department of Economics, University of California, Los Angeles, California

Hayne E. Leland, School of Business Administration, University of California, Berkeley, California

George Leitmann, Department of Mechanical Engineering, University of California, Berkeley, California

Tracy R. Lewis, Department of Economics, University of Arizona, Tucson, Arizona

Pan-Tai Liu, Mathematics Department, University of Rhode Island, Kingston, Rhode Island

Hajime Myoken, Faculty of Economics, Nagoya City University, Mizuhocho Mizuhoku, Nagoya, Japan

Richard Schmalensee, Sloan School of Management, Massachusetts Institute of Technology, Cambridge, Massachusetts

Martin Shubik, Cowles Foundation, Yale University, New Haven, Connecticut

Matthew J. Sobel, College of Industrial Management, Georgia Institute of Technology, Atlanta,

Jon G. Sutinen, Department of Resource Economics, University of Rhode Island, Kingston, Rhode Island

Pravin Varaiya, Department of Electrical Engineering and Computer Sciences, and the Electronics Research Laboratory, University of California, Berkeley, California

Jean Walrand, Department of Electrical Engineering and Computer Sciences, and the Electronics Research Laboratory, University of California, Berkeley, California. Present address: School of Electrical Engineering, Cornell University, Ithaca, New York

Henry Y. Wan Jr., Department of Economics, Cornell University, Ithaca, New York

Preface

As an outgrowth of the advancement in modern control theory during the past 20 years, dynamic modeling and analysis of economic systems has become an important subject in the study of economic theory. Recent developments in dynamic utility, economic planning, and profit optimization, for example, have been greatly influenced by results in optimal control, stabilization, estimation, optimization under conflicts, multicriteria optimization, control of large-scale systems, etc.

The great success that has been achieved so far in utilizing modern control theory in economic systems should be attributed to the effort of control theorists as well as economists. Collaboration between the two groups of researchers has proven to be most successful in many instances; nevertheless, the gap between them has existed for some time. Whereas a control theorist frequently sets up a mathematically feasible model to obtain results that permit economic interpretations, an economist is concerned more with the fidelity of the model in representing a real-world problem, and results that are obtained (through possibly less mathematical analysis) are due largely to economic insight.

The papers appearing in this volume are divided into three parts. In Part I there are five papers on the application of control theory to economic planning. Part II contains five papers on exploration, exploitation, and pricing of extractive natural resources. Finally, in Part III, some recent advances in large-scale systems and decentralized control appear.

These papers are contributions from control theorists and economists. Each paper presents its own perspective of future developments in dynamic economic theory. It is hoped that this collection will help stimulate interaction between the control-theoretic and the economically oriented approaches and that the volume will be of interest to researchers in applied mathematics, economics, management sciences, etc. It is also hoped that this collection will provide a periscopic view of some recent progress in mathematical economics.

In editing this volume we have been assisted by Dr. Tamer Başar, Professor Jon Sutinen, and Professor Henry Wan. Most of the papers in this volume were presented at the Third Kingston Conference on Differential Games and Control Theory, held in June 1978; the theme of the conference was dynamic optimization and mathematical economics. Financial support from the Office of Naval Research and the University of Rhode Island in organizing the conference are gratefully acknowledged. Finally, we would like to thank Professor George Leitmann and Professor Angelo Miele for making possible the publication of this volume in the present series.

Kingston, Rhode Island Pan-Tai Liu

Contents

PART I. CONTROL THEORY IN ECONOMIC SYSTEMS

1. Asset Pricing in an Economy with Production: A "Selective" Survey of Recent Work on Asset-Pricing Models 5
   William A. Brock
2. Alternative Long-Run Goals and the Theory of the Firm: Why Profit Maximization May Be a Better Assumption Than You Think . . . 31
   Hayne E. Leland
3. Mathematical Optimization and Economic Behavior 51
   Sheldon S. L. Chang
4. Performance Improvement of Uncertain Macroeconomic Systems . . 71
   George Leitmann and Henry Y. Wan Jr.
5. Stochastic Games, Oligopoly Theory, and Competitive Resource Allocation 89
   Martin Shubik and Matthew J. Sobel

PART II. DYNAMICAL THEORY OF THE ECONOMICS OF EXTRACTIVE NATURAL RESOURCES

6. Optimal Pricing, Use, and Exploration of Uncertain Natural Resource Stocks 105
   Kenneth J. Arrow and Sheldon S. L. Chang
7. Restricted Access to Common-Property Fishery Resources: A Game-Theoretic Analysis 117
   Colin W. Clark
8. Cartel and Oligopoly Pricing of Nonreplenishable Natural Resources 133
   Tracy R. Lewis and Richard Schmalensee
9. Optimum Extraction of an Exhaustible Resource: A Mathematical Analysis 157
   Pan-Tai Liu
10. Royalties and the Extraction Rate of an Exhaustible Resource. . . 169
    Jon G. Sutinen

PART III. LARGE-SCALE SYSTEMS AND DECENTRALIZED CONTROL

11. Dynamics and Control of a System Composed of a Large Number of Similar Subsystems 183
Masanao Aoki

12. Hierarchical Decisionmaking under Uncertainty 205
Tamer Başar

13. A Game-Theoretic Approach to Bureaucratic Behavior 223
Dagobert L. Brito and Michael D. Intriligator

14. Optimal Stabilization Policies for Decentralized Macroeconomic Systems with Conflicting Targets 237
Hajime Myoken

15. A Minimum Principle for Decentralized Stochastic Control Problems 253
Pravin Varaiya and Jean Walrand

INDEX 267

Part I

Control Theory in Economic Systems

The dynamic feature of economic systems has become increasingly important as the economic systems of modern society become more complex. The concept of state variables, which summarize the past history of a system, is essential for describing problems in economics, while control variables play the role of intertemporal decision making. Originally developed by engineers and mathematicians, control theory is now widely used in economics.

The importance of control theory in economics can be demonstrated by the extent of the literature that has been published, especially during the last ten years. Recent advances in economic planning, growth theory, and other aspects of the theory of firms have resulted from applications of modern control techniques.

The objectives in control of an economic system include optimization, stabilization, and a combination of the two. Optimization refers to maximization of one or more payoffs, such as profit flow and utility levels of capital assets at the end of the planned horizon. Stabilization means steering the state of the system (demand, price, laborers, or capital) toward a target level by proper choice of the control (investment, production, or fiscal policy).

The evolution of an economic system in general can be described by accumulation equations, which are equivalent to state equations in control theory. The objective functional is often the integral of an instantaneous utility function over a planned horizon. The problem is then analyzed by standard techniques in optimal control, e.g., the maximum principle and dynamic programming.

While quantitative results that bear some economic significance are desirable, they are not always obtainable. In many cases, qualitative interpretation of necessary conditions (for optimality) can be obtained with a moderate amount of mathematical analysis. A fundamental example is given by interpreting the adjoint variable in the maximum principle as the shadow price, which weighs future accumulation in the system against present utility. In general, a certain amount of economic insight is needed to "penetrate" a set of mathematical conditions and formulate some underlying economic theory.

A special feature of the control problem in economics is that the planned horizon may be infinite. In such a case, the transversality condition, which is part of the necessary condition for optimality, has to be formulated and utilized in the limiting sense as time goes to infinity. Steady-state solutions stand for some kind of dynamic equilibrium in the long run. They can be obtained (with the aid of transversality conditions) by setting all the derivatives with respect to time equal to zero. This is discussed in the papers by Chang and Leland.

Chang's paper illustrates that the standard optimal control theory can be applied to diverse economic problems. Three examples are given: (1) a comprehensive energy-planning problem for the government, (2) the optimal investment problem for a computer-using firm, in which the investment in research and development takes the form of developing programs, subroutines, etc., and (3) the optimal saving problem for the consumers, in which the impact of changing the initial capital value is analyzed in detail.

Leland's paper applies the conventional control model to a novel problem in the theory of firms. Suppose that real-life firms maximize some objective functions other than profit. Then for a large class of cases, the optimal decisions of the firm approach the optimal policies under profit maximization in the long run. This is in contrast to results from static models, which indicate that the decisions of a firm with alternative objectives differ from the decisions based on profit maximization.

The theory of finance deals with many aspects of a firm's performance as a corporation. One of the most important problems is the optimal investment and consumption plan over time. This includes (1) the amounts of investment and consumption at any time and (2) the allocation of investment among different types of assets.

Brock's paper contains a survey of his works and related works on asset pricing. Starting with an N-process stochastic growth model, described by stochastic difference equations, he characterizes optimal paths generated by optimum policies by means of the principle of optimality. If all consumers are assumed to be identical, their optimal consumption-investment decisions imply that production must be carried out in a certain fashion for a multigood model, buffeted by random shocks from nature. The profit and the stock price of the firms carrying out the production processes can then be derived. The limiting distribution of capital assets corresponds to the limiting distribution of stock prices.

For the one-consumer case, the N-process growth model is then converted into an asset-pricing model by introducing competitive rental markets for the capital goods and a market for pure rents generated by each individual firm. A general economic theory, based on the concept of a rational expectation equilibrium, is established and a unique asset-pricing function is indicated to exist.

In the modeling of an economic system, it is frequently necessary to consider uncertainties. One way to do so is to take the probabilistic approach of assuming that some parameters in the system are random variables or that the system is subject to some disturbances that can be described as stochastic processes. We then have a stochastic control problem in which the stochastic maximum principle or stochastic dynamic programming can be applied. For example, the fluctuation of the stock market can be described as a Brownian motion, and the accumulation equation for the capital of a firm then becomes an Ito's equation. Mathematical theory in control of dynamic systems described by Ito's equations has been fairly well established. The application of such a theory in economics is currently an area of active research.

Another way to describe uncertainties is to assume that they are bounded in some way but are otherwise unknown. The optimization problem is then formulated as one of finding the feedback control from within a certain class and minimizing the maximum values over all uncertain quantities. This worst-case, nonstochastic approach is often more natural and realistic, though the actual computation of a minimax control is generally complicated. The same principle can be applied to stabilization against uncertain disturbances by minimizing the maximum possible values of a Lyapunov function. The paper by Leitmann and Wan follows this approach in discussing the stabilization of a macroeconomic system that contains some unknown characteristics. They show that under some conditions, global, asymptotic stability can be guaranteed, uniformly over a class of bounded disturbances. If these conditions are absent, due either to observational error or to delay, or due to limitation of control instruments in their magnitude or scope, the performance of the economy can still be improved by adopting certain policies.

In large firms, pricing and resource allocation are among the most important aspects of decisionmaking. Shubik and Sobel define discrete time sequential games as multiperson Markov decision processes. They use such game models to describe dynamic oligopolistic market situations and other competitive resource allocation problems. In addition to addressing themselves to some issues that arise in such models, they discuss some principal sufficient conditions for optimality satisfied by various dynamic oligopoly models.

1

Asset Pricing in an Economy with Production: A "Selective" Survey of Recent Work on Asset-Pricing Models

WILLIAM A. BROCK

1. Introduction

This paper surveys an intertemporal general equilibrium theory of capital asset pricing. It is an attempt to put together ideas from the literatures on modern finance, stochastic growth models, and general equilibrium theory. In this way we shall obtain a theory capable of addressing general equilibrium questions such as the following: What is the impact of an increase in the corporate income tax on the relative prices of risky stocks? What is the impact of an increase in progressivity of the personal income tax on the relative price structure of risky assets? This paper discusses only recent literature that is closely related to my own work. Hence it should be read with this disclaimer in mind. Furthermore, because of space limitations, theorems and proofs will be loosely stated.

The theory presented here derives part of its inspiration from Merton (Ref. 1). However, Merton's intertemporal capital asset-pricing model (ICAPM) is not a general equilibrium theory in the sense of Arrow–Debreu. That is, the *technological* sources of uncertainty are not related to the equilibrium prices of the risky assets in Merton (Ref. 1). We do that here and preserve the empirical tractability of Merton's formulation.

WILLIAM A. BROCK • Department of Economics, University of Chicago, Chicago, Illinois, and Department of Economics, University of Wisconsin, Madison, Wisconsin.

One strand of the work surveyed here in the one-consumer case modifies the stochastic growth models of Brock and Mirman (Refs. 2, 3), Radner (Ref. 4), and Jeanjean (Ref. 5) in order to put a nontrivial investment decision into the asset-pricing model of Lucas (Ref. 6). Brock (Ref. 7) does this in a way that preserves the empirical tractability of the Merton formulation and at the same time determines the risk prices derived by Ross (Ref. 8) in his arbitrage theory of capital asset pricing. Ross's price of systematic risk k at date t, denoted by λ_{kt} and induced by the source of systematic risk δ_{kt}, is determined by the covariance of the marginal utility of consumption with λ_{kt}. In this way we discover assumptions about tastes and technology needed to generate Ross's λ_{kt} and show exactly how they are determined by the interaction of sources of production uncertainty and the demand for risky assets.

Prescott and Mehra (Ref. 9) insert an investment decision into Lucas's model (Ref. 6) and develop a dynamic programming recursive equilibrium framework that covers many recent models of asset pricing such as Brock (Ref. 7) and Lucas (Ref. 6).

Another line of work is developed in continuous time by Cox, Ingersoll, and Ross (Ref. 10), and others. This work seeks to combine intertemporal general equilibrium analysis with the pricing theory of derivative claims that culminated in the work of Black and Scholes (Ref. 11).

This paper proceeds as follows. Section 1 contains the introduction. Section 2 presents an N-process stochastic growth model, which forms the basis for the quantity side of the asset-pricing model developed in Section 3 for the one-consumer case. It will lead to useful insights for the multiconsumer case.

Section 2 describes optimum paths generated in the N-process model by time-independent continuous optimum policy functions a la Bellman. Sufficient conditions on tastes and technology are developed that allow the derivation of a functional equation that determines the state valuation function, with the use of methods adapted from a paper by Koopmans (Ref. 12). We also discuss the problem of finding sufficient conditions on tastes and technology such that for any initial state the optimum stochastic process of investment converges in distribution to a limit distribution independent of the initial state.

For the one-consumer case Section 3 converts the growth model of Section 2 into an asset-pricing model by introducing competitive rental markets for the capital goods and by introducing a market for claims to the pure rents generated by the ith firm, $i = 1, 2, \ldots, N$. Each of the N processes is identified with one "firm." Firms pay out rentals to consumers. The residual is pure rent. Paper claims to the pure rent generated by each firm i and a market for these claims are introduced along the line of

Lucas (Ref. 6), i.e., both sides of the economy possess subjective distributions on pure rents, capital rental rates, and share prices. Both sides draw up demand and supply schedules conditioned on their subjective distributions. Market clearing introduces an objective distribution on pure rents, capital rental rates, and share prices. A rational expectations equilibrium, abbreviated R.E.E., is defined by the requirement that the objective distribution equal the subjective distribution at each date.

Section 3 shows (for the one-consumer case), with the use of recent results of Benveniste and Scheinkman (Ref. 13), that the quantity side of an R.E.E. is identical to the quantity side of the N-process growth model developed in Section 2. The key idea used is the Benveniste–Scheinkman result that for utility functions that are additively separable over time, the standard transversality condition at infinity is necessary as well as sufficient for an infinite-horizon concave programming problem. This theory is used to characterize optimal plans by each consumer.

The financial side of the economy is now easy to develop. A unique asset-pricing function for stock i of the form $P_i(y)$ is indicated to exist by use of a contraction mapping argument along the line of Lucas (Ref. 6), where y describes the state of the economy (which may include its history).

Section 4 reviews some other asset-pricing models. The first model, developed by Becker (Ref. 14), is a deterministic version of the one-consumer model developed in Section 3. He shows that if each consumer's utility function is a discounted sum of period utilities with time-independent discount factor and time-independent period utility function, then the consumer that discounts the future the least eventually ends up with all the capital stock.

The next paper, by Magill (Ref. 15), applies the recent work of Bismut (Ref. 16) on continuous-time stochastic optimal control to economic dynamics. Magill's paper uses Bismut's work to characterize rational expectations equilibria in an economic world driven by Ito processes.

An important class of continuous-time stochastic asset-pricing models is reviewed in Section 5. These models, built by Cox, Ingersoll, and Ross (Ref. 10; hereafter CIR) are general equilibrium versions of Merton's intertemporal capital asset–pricing model (Ref. 1). CIR also add a class of derivative claims and develop generalized Black–Scholes (Ref. 11) partial differential equations for the prices of these claims.

Finally, Section 6 contains a summary.

2. The Optimal Growth Model

Since the optimal growth model is studied in detail in Brock (Ref. 7), we shall be brief where possible.

The model is given by the following:

Maximize

$$E_1 \sum_{t=1}^{\infty} \beta^{t-1} u(c_t), \tag{1}$$

subject to

$$c_{t+1} + x_{t+1} - x_t = \sum_{i=1}^{N} [g_i(x_{it}, r_t) - \delta_i x_{it}], \tag{2}$$

$$x_t = \sum_{i=1}^{N} x_{it}, \qquad x_{it} \geq 0, \qquad i = 1, 2, \ldots, N, t = 1, 2, \ldots, \tag{3}$$

$$c_t \geq 0, \qquad t = 1, 2, \ldots, \tag{4}$$

$$x_0, x_{i0} \quad (i = 1, 2, \ldots, N), r_0 \quad \text{historically given,} \tag{5}$$

where E_1, β, u, c_t, x_t, g_i, x_{it}, r_t, δ_i denote, respectively, mathematical expectation conditioned at time 1, discount factor on future utility, utility function of consumption, consumption at date t, capital stock at date t, production function of process i, capital allocated to process i at date t, random shock which is common to all processes i, and depreciation rate for capital installed in process i.

The space of $\{c_t\}_{t=1}^{\infty}$, $\{x_t\}_{t=1}^{\infty}$ over which the maximum is being taken in (1) needs to be specified. Obviously decisions at date t should be based only on information at date t. In order to make the choice space precise some formalism is needed. We borrow (copy) from Brock and Majumdar (Ref. 17) at this point.

The *environment* will be represented by a sequence $\{r_t\}_{t=1}^{\infty}$ of real vector-valued random variables which will be assumed to be independently and identically distributed. The common distribution of r_t is given by a measure $\mu: \mathscr{B}(R^m) \to [0, 1]$, where $\mathscr{B}(R^m)$ is the Borel σ-field of R^m. In view of a well-known one-to-one correspondence [see, e.g., Loeve (Ref. 18, pp. 230–231)], we can adequately represent the environment as a measure space $(\Omega, \mathscr{F}, \nu)$, where Ω is the set of all sequences of real m-vectors, $\mathscr{F}$ is the σ-field generated by cylinder sets of the form $\prod_{t=1}^{\infty} A_t$, where

$$A_t \in \mathscr{B}(R^m), \qquad t = 1, 2, \ldots,$$

and

$$A_t = R^m$$

for all but a finite number of values of t. Also ν (the stochastic law of the environment) is simply the product probability induced by μ (given the assumption of independence).

The random variables r_t may be viewed as the tth coordinate function on Ω, i.e., for any $\omega = \{\omega_t\}_{t=1}^{\infty} \in \Omega$, $r_t(\omega)$ is defined by

$$r_t(\omega) = \omega_t.$$

We shall refer to ω as a possible state of the environment (or an environment sequence) and to ω_t as the environment at date t. In what follows, $\mathscr{F}_t$ is the σ-field guaranteed by partial histories up to period t (i.e., the smallest σ-field generated by cylinder sets of the form $\prod_{\tau=1}^{\infty} A_\tau$ where A_τ is in $\mathscr{B}(R^m)$ for all t and $A_\tau = R^m$ for all $\tau > t$). The σ-field $\mathscr{F}_t$ contains all the information about the environment that is available at date t.

To express precisely the fact that decisions c_t, x_t depend only on information available when the decisions are made, we simply require that c_t, x_t be measurable with respect to $\mathscr{F}_t$.

Formally the maximization in (1) is taken over all stochastic processes $\{c_t\}_{t=1}^{\infty}$, $\{x_t\}_{t=1}^{\infty}$ that satisfy (2)–(5) and such that for each $t = 1, 2, \ldots$, c_t, x_t are measurable $\mathscr{F}_t$. Call such processes *admissible.*

Existence of an optimum $\{c_t\}_{t=1}^{\infty}$, $\{x_t\}_{t=1}^{\infty}$ may be established by imposing an appropriate topology $\mathscr{T}$ on the space of admissible processes such that the objective (1) is continuous in this topology and the space of admissible processes is $\mathscr{T}$-compact. While it is beyond the scope of this article to discuss existence, presumably a proof can be constructed along the lines of Bewley (Ref. 19).

The notation almost makes the working of the model self-explanatory. There are N different processes. At date t it is decided how much to consume and how much to hold in the form of capital. It is assumed that capital goods can be costlessly transformed into consumption goods on a one-for-one basis. After it is decided how much capital to hold, then it is decided how to allocate the capital across the N processes. After the allocation is decided, nature reveals the value of r_t, and $g_i(x_{it}, r_t)$ units of new production are available from process i at the end of period t. But $\delta_i x_{it}$ units of capital have evaporated at the end of t. Thus net new produce is $g_i(x_{it}, r_t) - \delta_i x_{it}$ from process i. The total produce available to be divided into consumption and capital stock at date $t + 1$ is given by

$$\sum_{i=1}^{N} [g_i(x_{it}, r_t) - \delta_i x_{it}] + x_t = \sum_{i=1}^{N} [g_i(x_{it}, r_t) + (1 - \delta_i)x_{it}]$$

$$\equiv \sum_{i=1}^{N} f_i(x_{it}, r_t) \equiv y_{t+1}, \tag{6}$$

where

$$f_i(x_{it}, r_t) \equiv g_i(x_{it}, r_t) + (1 - \delta_i)x_{it} \tag{7}$$

denotes the total amount of produce emerging from process i at the end of period t. The produce y_{t+1} is divided into consumption and capital stock at the beginning of date $t+1$, and so on.

Note that we are assuming that it is costless to install capital into each process i and that it is costless to allocate capital across processes at the beginning of each date t.

The objective of the optimizer is to maximize the expected value of the discounted sum of utilities over all consumption paths and capital allocations that satisfy (2)–(5).

To obtain sharp results we will place restrictive assumptions on this problem. We collect the basic working assumptions in one place.

Assumption 2.1. The functions $u(\cdot)$, $f_i(\cdot)$ are all concave, increasing, and twice continuously differentiable.

Assumption 2.2. The stochastic process $\{r_t\}_{t=1}^{\infty}$ is independently and identically distributed. Each $r_t:(\Omega, \mathcal{B}, \mu) \to R^m$, where $(\Omega, \mathcal{B}, \mu)$ is a probability space. Here Ω is the space of elementary events, $\mathcal{B}$ is the σ-field of measurable sets with respect to μ, and μ is a probability measure defined on subsets $B \subseteq \Omega$, $B \in \mathcal{B}$. Furthermore the range of r_t, $r_t(\Omega)$, is *compact.*

Assumption 2.3. For each $\{x_{i1}\}_{i=1}^{N}$, r_1 the problem (1) has a unique optimal solution (unique up to a set of realizations of $\{r_t\}$ of measure zero).

Notice that Assumption 2.3 is implied by Assumption 2.1 and strict concavity of u, $\{f_i\}_{i=1}^{N}$. Rather than try to find the weakest possible assumptions sufficient for uniqueness of solutions to (1), it seems simpler to reveal the role of uniqueness in what follows by simply assuming it. Furthermore since we are not interested here in the study of existence of optimal solutions, we have simply assumed existence also.

By Assumption 2.3 we see that to each output level y_t the optimum c_t, x_t, x_{it}, given y_t, may be written

$$c_t = g(y_t), \qquad x_t = h(y_t), \qquad x_{it} = h_i(y_t). \tag{8}$$

The optimum policy functions $g(\cdot)$, $h(\cdot)$, $h_i(\cdot)$ do not depend on t because the problem given by (1)–(5) is time stationary.

Another useful optimum policy function may be obtained. Given x_t and r_t, Assumption 2.3 implies that the optimal allocation $\{x_{it}\}_{i=1}^{N}$ and the next period's optimal capital stock x_{t+1} is unique. Furthermore these may be written in the form

$$x_{it} = a_i(x_t, r_{t-1}), \tag{9}$$

$$x_{t+1} = H(x_t, r_t). \tag{10}$$

Equations (9) and (10) contain r_{t-1} and r_t respectively because the allocation decision is made after r_{t-1} is known but before r_t is revealed, while the capital-consumption decision is made after y_{t+1} is revealed, i.e., after r_t is known.

Equation (10) looks very much like the optimal stochastic process studied by Brock and Mirman (Ref. 2) and by Mirman and Zilcha (Refs. 20–22). For the case $N = 1$ the stochastic difference equation (10) converges in distribution to a unique limit distribution independent of initial conditions (Refs. 2, 3). The same result may be obtained for our N-process model (Ref. 7). The following result gives conditions for necessity of the transversality condition at ∞.

Result 2.1. Assume Assumption 2.1. Also assume that units of utility may be chosen so that $u(c) \geq 0$, for all c. Furthermore assume that along optima

$$E_1\beta^{t-1}U(y_t) \to 0 \quad \text{as} \quad t \to \infty.$$

If $\{c_t\}_{t=1}^{\infty}$, $\{x_t\}_{t=1}^{\infty}$, $\{x_{it}\}_{i=1}^{N}$, $t = 1, 2, \ldots$, is optimal, then the following conditions must be satisfied:

For each i, t

$$u'(c_t) \geq \beta E_t\{u'(c_{t+1})f_i'(x_{it}, r_t)\}, \tag{10a}$$

$$u'(c_t)x_{it} = \beta E_t\{u'(c_{t+1})f_i'(x_{it}, r_t)x_{it}\}, \tag{10b}$$

$$\lim_{t\to\infty} E_1\{\beta^{t-1}u'(c_t)x_t\} = 0. \tag{10c}$$

Proof. See Brock (Ref. 7). □

Here $U(y)$ is the maximum of (1) given y.

2.1. The Price of Systematic Risk

Steve Ross (Ref. 8) produced a theory of capital asset pricing showing that the assumption that all systematic risk-free portfolios earn the risk-free rate of return plus the assumption that asset returns are generated by a K-factor model lead to the existence of "prices" λ_0, λ_1, $\lambda_2, \ldots, \lambda_K$ on mean returns and on each of the K factors. These prices satisfied the property that the expected return $E\tilde{\mathscr{Z}}_i \equiv a_i$ on each asset i was a linear function of the standard deviation of the returns on asset i with respect to each factor k, i.e.,

$$a_i = \lambda_0 + \sum_{k=1}^{K} \lambda_k b_{ki}, \qquad i = 1, 2, \ldots, N, \tag{11}$$

where the original model of asset returns is given by

$$\tilde{\mathfrak{X}}_i = a_i + \sum_{k=1}^{K} b_{ki}\tilde{\delta}_k + \tilde{\epsilon}_i, \qquad i = 1, 2, \ldots, N. \tag{12}$$

Here $\tilde{\mathfrak{X}}_i$ denotes random *ex ante* anticipated returns from holding the asset one unit of time, $\tilde{\delta}_k$ is systematic risk emanating from factor k, $\tilde{\epsilon}_i$ is unsystematic risk specific to asset i, and a_i, b_{ki} are constants. Assume that the means of $\tilde{\delta}_k$, $\tilde{\epsilon}_i$ are zero for each k, i, that $\tilde{\epsilon}_1, \ldots, \tilde{\epsilon}_N$ are independent, and that $\tilde{\delta}_k$, $\tilde{\epsilon}_i$ are uncorrelated random variables with finite variances for each k, i.

However Ross's model, like the standard capital asset-pricing models in finance, does not link the asset returns to underlying sources of uncertainty. Our growth model will be used as a module in the construction of an intertemporal general equilibrium asset-pricing model, where relationships of the form (12), and hence the $\lambda_0, \lambda_1, \ldots, \lambda_K$, are determined within the model. Such a model of asset price determination preserves the beauty and empirical tractability of the Ross–Sharpe–Lintner formulation but at the same time gives a context in which we can ask general equilibrium questions such as; What is the impact of an increase of the progressivity of the income tax on the demand for and supply of risky assets and the $\lambda_0, \lambda_1, \ldots, \lambda_K$?

Let us get on with relating the growth model to (11). For simplicity assume that all processes i are active, i.e., (10a) holds with equality. We repeat (10a) here for convenience:

$$u'(c_t) = \beta E_t\{u'(c_{t+1})f_i'(x_{it}, r_t)\}. \tag{13}$$

Now (12) is a special hypothesis about asset returns. What kind of hypothesis about "technological" uncertainty corresponds to (12)? As an example, for each $i = 1, 2, \ldots, N$ put

$$\begin{aligned} f_i(x_{it}, r_t) &\equiv (A_{it}^0 + A_{it}^1\tilde{\delta}_{1t} + A_{it}^2\tilde{\delta}_{2t} + \cdots + A_{it}^K\tilde{\delta}_{Kt})f_i(x_{it}) \\ &\equiv r_{it}f_i(x_{it}), \end{aligned} \tag{14}$$

where

$$A_{it}^k \equiv A_i^k$$

are constants and

$$\{\tilde{\delta}_{kt}\}_{t=1}^{\infty}$$

are independent and identically distributed random variables for each k and where for each k, t the mean of $\tilde{\delta}_{kt}$ is zero, the variance is finite, and $\tilde{\delta}_{st}$ is independent of $\tilde{\delta}_{kt}$ for each s, k, t. Furthermore assume that $f(\cdot)$ is concave, increasing, and twice differentiable, that $f'(0) = +\infty$, that $f'(\infty) = 0$, and that there is a bound ϵ_0 such that

$$r_{it} > \epsilon_0 > 0$$

with probability one for all i, t. These assumptions are stronger than necessary, but they enable us to avoid concern with technical tangentialities. For all t define

$$\tilde{\delta}_{0t} \equiv 1,$$

so that we may sum from $k = 0$ to K in (15).

Insert (14) into (15) to get for all t, k, i,

$$u'(c_t) = \beta E_t\left\{u'(c_{t+1})\left(\sum_{k=0}^{K} A_{it}^k\tilde{\delta}_{kt}\right)f_i'(x_{it})\right\}$$

$$= \sum_{k=0}^{K} ([A_{it}^k f_i'(x_{it})]E_t\{\beta u'(c_{t+1})\tilde{\delta}_{kt}\}). \tag{15}$$

Now set (15) aside for a moment, and look at the marginal benefit of saving one unit of capital and assigning it to process i at the beginning of period t. At the end of period t, r_t is revealed and extra produce

$$\tilde{\mathscr{X}}_{it} \equiv A_{it}^0 f_i'(x_{it}) + \sum_{k=1}^{K} A_{it}^k f_i'(x_{it})\tilde{\delta}_{kt} \tag{16}$$

emerges. Putting

$$a_i \equiv A_{it}^0 f_i'(x_{it}), \qquad b_{ki} \equiv A_{it}^k f_i'(x_{it}), \qquad \tilde{\delta}_{kt} = \tilde{\delta}_k, \tag{17}$$

we find that equation (16) is identical to Ross's (12) with $\tilde{\epsilon}_i \equiv 0$. We proceed now to generate the analogue to (11) in our model. Rewrite (15) using (17); thus

$$u'(c_t) = \sum_{k=1}^{K} b_{ki}E_t\{\beta u'(c_{t+1})\tilde{\delta}_{kt}\} + a_i E_t\{\beta u'(c_{t+1})\}. \tag{18}$$

Hence

$$a_i = \frac{u'(c_t)}{\beta E_t\{u'(c_{t+1})\}} - \sum_{k=1}^{K} b_{ki}\frac{E_t\{u'(c_{t+1})\tilde{\delta}_{kt}\}}{E_t\{u'(c_{t+1})\}}, \tag{19}$$

so that $\lambda_0, \lambda_1, \ldots, \lambda_K$ defined by

$$\lambda_0 \equiv \frac{u'(c_t)}{\beta E_t\{u'(c_{t+1})\}}, \qquad \lambda_k \equiv \frac{-E_t\{u'(c_{t+1})\tilde{\delta}_{kt}\}}{E_t\{u'(c_{t+1})\}} \tag{20}$$

yield

$$a_i = \lambda_0 + \sum_{k=1}^{K} b_{ki}\lambda_k. \tag{21}$$

Here t subscripts are dropped for simplicity.

These results are extremely suggestive and show that the model studied in this section may be quite rich in economic content. Although the model is normative, in the next section we shall turn it into an equilibrium asset-pricing model so that the λ_k become equilibrium risk prices. Let us explore the economic meanings of (20) in some detail.

Suppose that $K = 1$ and that there is a risk-free asset N in the sense that

$$b_{N1} \equiv A^1_{Nt} f'(x_{Nt}) = 0, \tag{22}$$

i.e.,

$$A^1_{Nt} = 0. \tag{23}$$

Then by (23)

$$a_N = \lambda_0, \qquad a_i = a_N + b_{1i}\lambda_1, \tag{24}$$

so that for all $i, j \neq N$,

$$(a_i - a_N)/b_{1i} = (a_j - a_N)/b_{1j}. \tag{25}$$

The second part of equation (24) corresponds to the security market line that says that expected return and risk are linearly related in a one-factor model. Equation (25) corresponds to the usual Sharpe–Lintner–Mossin result of the capital asset–pricing model that in equilibrium the "excess return" per unit of risk must be equated across all assets.

The economic interpretation of λ_0 given in (20) is well known and needs no explanation here. Look at the formula for λ_k. The covariance of the marginal utility of consumption at time $t+1$ with the zero-mean finite-variance shock $\tilde{\delta}_{kt}$ appears in the numerator. Output increases when $\tilde{\delta}_{kt}$ increases and

$$c_{t+1} = g(y_{t+1})$$

does not decrease when y_{t+1} increases; therefore this covariance is likely to be negative so the sign of λ_k is positive. We shall look into the determinants of the magnitudes of $\lambda_0, \lambda_1, \ldots, \lambda_K$ in more detail later. Let us show how this model may be helpful in the empirical problem in estimating the $\lambda_0, \lambda_1, \ldots, \lambda_K$ from time series data.

First how is one to close Ross's model (12), since the $\tilde{\mathscr{X}}_i$ are subjective? The most natural way to close the model in markets as well organized as U.S. securities markets would seem to be rational expectations: The subjective distribution of $\tilde{\mathscr{X}}_i$ is equal to the actual or objective distribution of $\tilde{\mathscr{X}}_i$. We shall show that our asset-pricing model under rational expectations generates the same solution as the normative model discussed before. Hence the convergence theorem implies that $\{x_t, c_t, x_{1t}, x_{2t}, \ldots, x_{Nt}\}_{t=1}^{\infty}$ converges to a stationary stochastic process.

Thus the mean ergodic theorem, which says very loosely that the time average of any function G of a stationary stochastic process equals the average of G over the stationary distribution of that process, allows us to apply time series methods developed for stationary stochastic processes to estimate $\lambda_0, \lambda_1, \ldots, \lambda_K$.

3. Asset-Pricing Model

In this section we reinterpret the model of Section 2 and add to it a market for claims to pure rents so that it describes the evolution of the equilibrium context in which to discuss the martingale property of capital asset prices. This model also contains a nontrivial investment decision, a nontrivial market for claims to pure rents (i.e., a stock market), as well as a market for the pricing of the physical capital stock.

I believe that there is considerable benefit in showing how to turn optimal growth models into asset-pricing models. This is so because there is a large literature on stochastic growth models which may be carried over to the asset-pricing problem with little effort. Although the model presented here is somewhat artificial, studying it will yield techniques that can be used to study less artificial models.

We will build an asset-pricing model much like that of Lucas (Ref. 6). The model contains one representative consumer whose preferences are identical to the planner's preferences given in (1). The model contains N different firms who rent capital from the consumption side at rate $R_{i,t+1}$ at each date in order to maximize

$$\pi_{i,t+1} \equiv f_i(x_{it}, r_t) - R_{i,t+1}x_{it}. \tag{26}$$

Notice that it is assumed that each firm i makes its decision to hire x_{it} *after* r_t is revealed. Here $R_{i,t+1}$ denotes the rental rate on capital prevailing in industry i at date $t + 1$. It is to be determined within the model. These rather artificial "rental markets" are introduced in order to obtain Lemma 3.2.

The model introduces a stock market in such a way that the real-quantity side of the model is the same as that of the growth model in equilibrium. The model is closed under the assumption of rational expectations. The quantity side of the model is essentially an Arrow–Debreu model, as is the model of Lucas (Ref. 6), i.e., we introduce securities markets in a way that assures enough securities so that any equilibrium is a Pareto optimum. However, there is a separate market in which claims to the rents (26) are competitively traded. In Arrow–Debreu the rents are redistributed in a lump sum fashion.

The model is in the spirit of Lucas's model, where each firm i has outstanding one perfectly divisible equity share. Ownership of $\alpha\%$ of the equity shares in firm i at date t entitles one to $\alpha\%$ of profits of the firm i

at date $t + 1$. Equilibrium asset prices and equilibrium consumption, capital, and output are determined by optimization under the hypothesis of rational expectations much as in Lucas (Ref. 6).

3.1. The Model

There is one representative consumer (or a "representative stand-in," as Lucas calls him) that is assumed to solve the following problem:

Maximize

$$E_1 \sum_{t=1}^{\infty} \beta^{t-1} u(c_t) \tag{27}$$

subject to

$$c_t + x_t + P_t \cdot Z_t \leq \pi_t \cdot Z_{t-1} + P_t \cdot Z_{t-1} + \sum_{i=1}^{N} R_{it} x_{i,t-1} \equiv y_t, \tag{28}$$

$$c_t \geq 0, \quad x_t \geq 0, \quad Z_t \geq 0, \quad x_{it} \geq 0, \quad i = 1, 2, \ldots, N, \quad \text{all } t, \tag{29}$$

$$c_1 + x_1 + P_1 \cdot Z_0 \leq \pi_1 \cdot Z_0 + P_1 \cdot Z_0 + \sum_{i=1}^{N} R_{i1} x_{i0} \equiv y_1, \tag{30}$$

given

$$Z_0 \equiv 1, \quad R_{i1} \equiv f_i'(x_{i0}, r_0), \quad \pi_{i1} \equiv f_i(x_{i0}, r_0) - f_i'(x_{i0}, r_0) x_{i0}, \quad x_0, \quad \{x_{i0}\}_{i=1}^{N},$$

where c_t, x_t, P_{it}, Z_{it}, π_{it}, R_{it}, all assumed measurable $\mathcal{F}_t$, denote consumption at date t, total capital stock owned at date t by the consumer, price of one share of firm i at date t, number of shares of firm i owned by the individual at date t, profits of firm i at date t, and rental factor (i.e., $R_{it} \equiv$ principal plus interest) obtained on a unit of capital leased to firm i. Here the centered dot denotes scalar product.

Firm i is assumed to hire x_{it} in order to maximize (26). The consumer is assumed to lease capital x_{it} at date t to firm i before r_t is revealed. Hence $R_{i,t+1}$ is uncertain at date t. The consumer, in order to solve his problem at date 1, must form expectations on $\{P_{it}\}_{t=1}^{\infty}$, $\{R_{it}\}_{t=1}^{\infty}$, $\{\pi_t\}_{t=1}^{\infty}$ and must maximize (27) subject to (28)–(30). In this way notional demands for consumption goods and equities as well as notional supplies of capital stocks and capital services to each of the N firms are drawn up by the consumer side of the economy. Similarly for the firm side. We close the model with the following definition.

Definition 3.1. The stochastic process $\mathcal{R} \equiv \langle\{\bar{P}_{it}\}_{t=1}^{\infty}, \{\bar{R}_{it}\}_{t=1}^{\infty}, \{\bar{\pi}_{it}\}_{t=1}^{\infty};$ $\{\bar{x}_{it}\}_{t=1}^{\infty}, \{\bar{Z}_{it}\}_{t=1}^{\infty}, i = 1, 2, \ldots, N, \{\bar{c}_t\}_{t=1}^{\infty}, \{\bar{x}_t\}_{t=1}^{\infty}\rangle$ is a *rational expectations equilibrium* (R.E.E.) if the consumer facing $\mathcal{P} \equiv \langle\{\bar{P}_{it}\}_{t=1}^{\infty}, \{\bar{R}_{it}\}_{t=1}^{\infty},$ $\{\bar{\pi}_{it}\}_{t=1}^{\infty}\rangle$ chooses

$$x_t = \bar{x}_t, \quad x_{it} = \bar{x}_{it}, \quad c_t = \bar{c}_t, \quad Z_{it} = \bar{Z}_{it} \qquad \text{a.e.,} \tag{31}$$

and the ith firm chooses

$$x_{it} = \bar{x}_{it} \tag{32}$$

and furthermore

$$\text{(asset market clears)} \qquad \bar{Z}_{it} \leq 1, \qquad \text{if } \bar{Z}_{it} < 1, \bar{P}_{it} = 0 \qquad \text{a.e.,} \tag{33}$$

$$\text{(goods market clears)} \quad \bar{c}_t + \bar{x}_t = \sum_{i=1}^{N} f_i(\bar{x}_{i,t-1}, r_{t-1}) \qquad \text{a.e.,} \tag{34}$$

$$\text{(capital market clears)} \quad \sum_{i=1}^{N} \bar{x}_{it} = \bar{x}_t \qquad \text{a.e.} \tag{35}$$

(Here a.e. means almost everywhere.) This ends the definition of R.E.E. that we will use in this paper.

It is easy to write down first-order necessary conditions for an R.E.E. Let us start on the consumer side first. We drop the overbars for simplicity. At date t if the consumer buys a share of firm i, the cost is P_{it} units of consumption goods. The marginal cost at date t in utils foregone is $u'(c_t)P_{it}$. At the end of period t, r_t is revealed and $P_{i,t+1}$, $\pi_{i,t+1}$ become known. Hence the consumer obtains

$$u'(c_{t+1})(P_{i,t+1} + \pi_{i,t+1}) \tag{36}$$

extra utils at the beginning of $t + 1$ if he collects $\pi_{i,t+1}$ and sells the share "exdividend" at $P_{i,t+1}$. But these utils are uncertain and are received one period into the future. The expected present value of utility gained at $t + 1$ is

$$\beta E_t\{u'(c_{t+1})(P_{i,t+1} + \pi_{i,t+1})\}. \tag{37}$$

Consumer equilibrium in the market for asset i requires that the marginal opportunity cost at date t be greater than or equal to the present value of the marginal benefit of dividends and the exdividend sale price at date $t + 1$:

$$P_{it}u'(c_t) \geq \beta E_t\{u'(c_{t+1})(\pi_{i,t+1} + P_{i,t+1})\} \qquad \text{a.e.,} \tag{38a}$$

$$P_{it}u'(c_t)Z_{it} = \beta E_t\{u'(c_{t+1})(\pi_{i,t+1} + P_{i,t+1})\}Z_{it} \qquad \text{a.e.} \tag{38b}$$

Similar reasoning in the rental market yields

$$u'(c_t) \geq \beta E_t\{u'(c_{t+1})(R_{i,t+1})\} \qquad \text{a.e.} \tag{39a}$$

$$u'(c_t)x_{it} = \beta E_t\{u'(c_{t+1})R_{i,t+1}\}x_{it} \qquad \text{a.e.} \tag{39b}$$

It would be nice if the first-order necessary conditions (38), (39) characterized consumer optima. But it is well known that a "transversality condition" at infinity (TVC_∞) is also needed to completely characterize optima. Recent work by Benveniste and Scheinkman (Ref. 13) allows us to prove the following result.

Lemma 3.1. Assume Assumption 2.1. Furthermore assume that $\mathcal{P}$ is such that

$$W(y_t, t) \to 0, \qquad t \to \infty,$$

where $W(y_t, t)$ is defined by

$$W(y_t, t) = \text{Maximum } E_1 \sum_{s=t}^{\infty} \beta^{s-1} u(c_s)$$

subject to (28)–(30) with t replaced by s and 1 replaced by t. Here y_t denotes the right-hand side of (28). Then given $\{P_{it}\}_{t=1}^{\infty}$, $\{\pi_{it}\}_{t=1}^{\infty}$, $\{R_{it}\}_{t=1}^{\infty}$, $i = 1, 2, \ldots, N$, optimum solutions $\{Z_{it}\}_{t=1}^{\infty}$, $\{x_{it}\}_{t=1}^{\infty}$, $i = 1, 2, \ldots, N$, $\{c_t\}_{t=1}^{\infty}$, $\{x_t\}_{t=1}^{\infty}$ to the consumers' problem (27) subject to (28)–(30) are characterized by (38)–(39) and

$$\text{TVC}_\infty(\text{equity market}) \qquad \lim_{t\to\infty} E_1\{\beta^{t-1} u'(c_t) P_t \cdot Z_t\} = 0, \tag{40}$$

$$\text{TVC}_\infty(\text{capital market}) \qquad \lim_{t\to\infty} E_1\{\beta^{t-1} u'(c_t) x_t\} = 0. \tag{41}$$

Proof. See Brock (Ref. 7). □

Remark 3.1. The assumption that $E_1 W(y_t, t) \to 0$ as $t \to \infty$ restrains $\mathcal{P}$. It requires that $\mathcal{P}$ be such that along any path in $\mathcal{P}$ utils cannot grow faster than β^t on the average. A general sufficient condition on $\mathcal{P}$ for $E_1 W(y_t, t) \to 0$ can be given by what should be a straightforward extension of the methods of Brock and Gale (Ref. 23) and McFadden (Ref. 24) to our setup.

An obvious sufficient condition is that the utility function be bounded, i.e., there are numbers $B < \bar{B}$ such that for all $c \geq 0$,

$$B \leq u(c) \leq \bar{B}.$$

Remark 3.2. The method used here of introducing a stock market into this type of model where an investment decision is present was first developed by Scheinkman (Ref. 25) in the certainty case.

The following is a basic lemma.

Lemma 3.2. (i) Let $X = \langle\{\bar{c}_t\}_{t=1}^{\infty}, \{\bar{x}_{it}\}_{t=1}^{\infty}, \{\bar{x}_t\}_{t=1}^{\infty}\rangle$ solve the optimal growth problem (1). Define

$$\bar{R}_{it+1} \equiv f_i'(\bar{x}_{it}, r_t), \qquad \bar{\pi}_{i,t+1} \equiv f_i(\bar{x}_{it}, r_t) - f_i'(\bar{x}_{it}, r_t)\bar{x}_{it}. \tag{42}$$

Then construct $\{\bar{P}_{it}\}_{t=1}^{\infty}$, $i = 1, 2, \ldots, N$ to satisfy the consumer's first-order necessary conditions of optimality and the consumer's transversality condition at infinity. Put

$$\bar{Z}_{it} = 1. \tag{43}$$

Then $\langle\{\bar{P}_{it}\}_{t=1}^{\infty}, \{\bar{R}_{it}\}_{t=1}^{\infty}, \{\bar{\pi}_{it}\}_{t=1}^{\infty}, \{\bar{x}_{it}\}_{t=1}^{\infty}, \{\bar{Z}_{it}\}_{t=1}^{\infty}, i = 1, 2, \ldots, N, \{\bar{c}_t\}_{t=1}^{\infty}, \{\bar{x}_t\}_{t=1}^{\infty}\rangle \equiv \mathcal{R}$ is an R.E.E.

(ii) Let $\mathcal{R}$ be an R.E.E. Then X solves the optimal growth problem (1).

Proof. See Brock (Ref. 7). □

The discussion of the relationship between the growth model of Section 2 and the risk prices of Ross will facilitate the economic interpretation of an R.E.E. stochastic process

$$\{\bar{R}_{it}\}_{t=1}^{\infty}, \qquad \{\bar{P}_{it}\}_{t=1}^{\infty}, \qquad \{\bar{\pi}_{it}\}_{t=1}^{\infty}.$$

For simplicity, drop the overbars from equilibrium quantities from this point on. Assume that conditions are such that all asset prices are positive with probability one in equilibrium. Then $\bar{Z}_{it} = 1$ with probability one, and from (38b) we get for each t

$$u'(c_t) = \beta E_t\{u'(c_{t+1})\mathcal{Z}_{it}\}, \qquad \mathcal{Z}_{it} \equiv (P_{i,t+1} + \pi_{i,t+1})/P_{it}. \tag{44}$$

Now because profit maximization implies that $f_i'(x_{it}, r_t) = R_{i,t+1}$, we have

$$\pi_{i,t+1} = f_i(x_{it}, r_t) - f_i'(x_{it}, r_t)x_{it}. \tag{45}$$

Turning to the rental market, suppose that all processes are used with probability one. Then (45) and (46) give for each i, t

$$u'(c_t) = \beta E_t\{u'(c_{t+1})f_i'(x_{it}, r_t)\}. \tag{46}$$

Examine the specification

$$f_i(x_{it}, r_t) = \left(\sum_{k=0}^{K} A_{it}^k \tilde{\delta}_{kt}\right) f_i(x_{it}) \equiv r_{it} f_i(x_{it}) \tag{47}$$

developed in Section 2. Now we have

$$u'(c_t) = \sum_{k=1}^{K} b_{kit} E_t\{\beta u'(c_{t+1})\tilde{\delta}_{kt}\} + a_{it} E_t\{\beta u'(c_{t+1})\}$$

$$= \beta E_t\{u'(c_{t+1})\tilde{\mathcal{Z}}_{it}\}. \tag{48}$$

We are not entitled to write returns $\mathscr{X}_{it}$ defined by (44) in the linear Ross form (12) unless $P_i(y_{t+1})$ is linear in y_{t+1} even for the specification (47). An example is presented in Brock (Ref. 7) where $P_i(y_{t+1})$ turns out to be linear in y_{t+1}. But first we must show that an asset-pricing function exists.

3.2. Existence of an Asset-Pricing Function

In equilibrium the quantity side of the asset-pricing model is the same as the N-process growth model; therefore we may use the facts collected in Section 2 about the N-process growth model to prove the existence of an asset-pricing function $P(y)$ in much the same way that Lucas did (Ref. 6).

Assumption 3.1. Assume for all $r \in R$ that

(a) $f_i'(0, r) = +\infty$, $i = 1, 2, \ldots, N$,
(b) $\pi_i(x, r) \equiv f_i(x, r) - f_i'(x, r)x > 0$ for all $x > 0$.

Assumption 3.1a implies that (39a) holds with equality in equilibrium. Also Assumption 3.1b implies that (38a) holds with equality in equilibrium. Let us search, as does Lucas (Ref. 6), for a bounded continuous function $P_i(y)$ such that in equilibrium

$$P_{it}u'(c_t) = P_i(y_t)u'(c_t) = \beta E_t\{u'(c_{t+1})(\pi_{i,t+1} + P_i(y_{t+1}))\}. \tag{49}$$

Convert the foregoing problem into a fixed-point problem. Note first from Section 2 that

$$u'(c_t) = U'(y_t), \qquad t = 1, 2, \ldots, \tag{50}$$

$$\begin{aligned}\pi_{i,t+1} &= f_i(x_{it}, r_t) - f_i'(x_{it}, r_t)x_{it} \equiv \pi_i(x_{it}, r_t) = \pi_i(\eta_i(x_t)x_t, r_t)\\ &= \pi_i[\eta_i(h(y_t))h(y_t), r_t] \equiv J_i(y_t, r_t),\end{aligned} \tag{51}$$

$$y_{t+1} = \sum_{j=1}^{N} f_j(\eta_j(x_t)x_t, r_t) = \sum_{j=1}^{N} f_j[\eta_j(h(y_t))h(y_t), r_t] \equiv Y(y_t, r_t). \tag{52}$$

Put

$$G_i(y_t) \equiv \beta \int_{r\in R} U'[Y(y_t, r)]J_i(y_t, r)\mu(dr), \tag{53}$$

$$F_i(y_t) \equiv P_i(y_t)U'(y_t), \tag{54}$$

$$(T_iF_i)(y_t) \equiv G_i(y_t) + \beta \int_R F_i[Y(y_t, r)]\mu(dr). \tag{55}$$

Then for each i (49) may be written as

$$F_i(y_t) = (T_iF_i)(y_t). \tag{56}$$

Problem (56) is a fixed-point problem in that we search for a function F_i that remains fixed under operator T_i. To use the contraction mapping theorem to find a fixed point F_i, we must first show that T_i sends the class of bounded continuous functions on $[0, \infty)$, call it $C[0, \infty)$, into itself. The results of Section 2 established that all the functions listed in (50)–(55) are continuous in y_t. We need the following lemma.

Lemma 3.3. If $U(y)$ is bounded on $[0, \infty)$, then $G_i(y)$ is bounded.

Proof. See Brock (Ref. 7). □

We must show that if

$$\|F_i\| \equiv \sup_{y \in [0, \infty)} |F_i(y)| \tag{57}$$

is chosen to be the norm on $C[0, \infty)$, then T_i is a contraction with modulus β. It is well known that $C[0, \infty)$ endowed with this norm is a Banach space.

Lemma 3.4. $T_i : C[0, \infty) \to C[0, \infty)$ is a contraction with modulus β.

Proof. See Brock (Ref. 7). □

Theorem 3.1. For each i there exists exactly one asset-pricing function of the form $P_i(y)$ where $P_i \in C[0, \infty)$.

Proof. Apply the contraction mapping theorem to produce a fixed point $\bar{F}_i(y) \in C[0, \infty)$. Put

$$P_i(t) = \bar{F}_i(y)/U'(y). \tag{58}$$

It is clear that $P_i(y)$ satisfies (49). Furthermore by the very definition of T_i any $P_i(y)$ that satisfies (49) is such that $P_i(\cdot)U'(\cdot) \equiv \bar{F}_i(\cdot)$ is a fixed point of T_i. □

Remark 3.3. Assumption 3.1a is not needed for the existence theorem. Assumption 3.1b is needed in the theorem so that (49) holds with equality.

Remark 3.4. There may be equilibrium processes $\{P_{it}\}_{t=1}^{\infty}$, $i = 1, 2, \ldots, N$, other than those that may be written in the form $P_{it} = P_i(y_t)$ for some continuous bounded function $P_i(y)$.

4. Some Other Asset-Pricing Models

This section will be more speculative than the previous sections because we shall use the model of asset pricing presented before as an organizing vehicle in order to review some recent work by economists in economic dynamics. The deterministic models to be reviewed are to be related to the preceding stochastic model by replacing each random variable by its mean.

Robert Becker (Ref. 14) has constructed deterministic versions of the model with many capital goods where consumers have preferences of the form

$$U^h(c^h) = \sum_{t=1}^{\infty} \beta_h^{t-1} u_h(c_t) \tag{59}$$

and where separate markets for claims to the residuals $\{\pi_{i,t+1}\}$ are absent. Rental markets for capital goods are introduced by Becker as in the preceding model. Profits are redistributed lump sum as in the Arrow–Debreu private ownership economy. Under general conditions Becker proves existence of rational expectations equilibrium in the deterministic case. He characterizes consumers' optima, i.e., their intertemporal demand functions, as we did in Section 3.

Since Becker's technology as well as his utility functions is time stationary, he is therefore able to examine steady states. If borrowing is not allowed, he shows that a steady state exists and that the person with the largest β_h is the only one who owns any capital in steady state.

I believe that this type of result was first noticed by H. Uzawa. Assume that consumers are not allowed to borrow. Then it follows directly from Section 3 that in steady state

$$u_h'(\bar{c}^h) \geq \beta_h u_h'(\bar{c}^h)\bar{R}_i, \quad \bar{R}_i = f_i'(x_i'), \quad \sum_{i=1}^{N} \bar{x}_i = \bar{x}, \qquad h = 1, 2, \ldots, H, \quad i = 1, 2, \ldots, N, \tag{60}$$

must hold. Hence

$$1/\beta_h \geq f_i'(\bar{x}_i), \tag{61}$$

with equality for $\bar{x}_i^h > 0$. Thus if $0 < \beta_1 < \beta_2 < \cdots < \beta_H$, then (61) implies

$$1/\beta_1 > 1/\beta_2 > \cdots > 1/\beta_H \geq f_i'(\bar{x}_i). \tag{62}$$

If strict inequality holds in (62), then $\bar{x}_i^h = 0$ for all i, h. Hence $\bar{x} = 0$. If we assume that $f_i'(0) = +\infty$ for some i and that $u_h'(0) = +\infty$ for each h, then it is easy to follow Becker (Ref. 14) and show that $\bar{x} > 0$ for any steady state. Hence $\bar{x}_i^H > 0$ for some i so that $\bar{x}_i = \bar{x}_i^H$, $\bar{x} = \sum_{i=1}^N \bar{x}_i^H$, and

$$1/\beta_H = f_i'(\bar{x}_i), \qquad \text{for } i \text{ such that } \bar{x}_i > 0. \tag{63}$$

The result that the person with the lowest discount factor on future utility ends up with all the capital stock is troubling. On one hand, the result is rigid and does not seem to correspond to reality; on the other, we do not want to give up the ease of working with utility functions of the form $\sum_{t=1}^\infty \beta_h^{t-1} u_h(c_t^h)$. Some analysts insert capital directly into the utility function in order to avoid this all-or-none problem, but inserting capital into the utility function has been criticized as double counting. In any event it is not clear what service that capital is providing beyond its flow of income to earn it a place in $u_h(c_h)$. Later we shall make some speculative comments about other utility functions.

A utility function that is analytically tractable and will permit nondegenerate steady states is

$$U^h(c^h) \equiv u_h(c_1^h) + \beta_h(c_1^h)u_h(c_2^h) + \beta_h(c_1^h)\beta_h(c_2^h)u_h(c_3^h) + \cdots. \tag{64}$$

In this case the discrete-time Euler inequality of Section 3 becomes after cancellation

$$\begin{aligned} &u_h'(c_t^h) + \beta_h'(c_t^h)u_h(c_{t+1}^h) + \beta_h'(c_t^h)\beta_h(c_{t+1}^h)u_h(c_{t+2}^h) + \cdots \\ &\quad \geq [\beta_h(c_t^h)u_h'(c_{t+1}^h) + \beta_h(c_t^h)\beta_h'(c_{t+1}^h)u_h(c_{t+2}^h) + \cdots]f_i'(x_{it}), \end{aligned} \tag{65}$$

with equality if $x_{it}^h > 0$. In steady state (65) simplifies to

$$1/\beta_h(\bar{c}^h) \geq f_i'(\bar{x}_i), \tag{66}$$

with equality if $\bar{x}_i^h > 0$. The complete set of equations and inequalities that are necessary (we do not say "sufficient" since U^h may not be concave) for a steady state are, in addition to (66);

$$\bar{c}^h + \bar{x}^h = \sum_{i=1}^N \alpha_{hi}\bar{\pi}_i + \sum_{i=1}^N \bar{f}_i(\bar{x}_i)\bar{x}_i^h, \qquad \bar{x}^h \equiv \sum_{i=1}^N \bar{x}_i^h, \tag{67}$$

$$\bar{\pi}_i \equiv f_i(\bar{x}_i) - f_i'(\bar{x}_i)\bar{x}_i, \tag{68}$$

$$\bar{x} = \sum_{i=1}^N \sum_{h=1}^H \bar{x}_i^h, \qquad \bar{x}_i = \sum_{h=1}^H \bar{x}_i^h, \qquad \sum_{h=1}^H \bar{c}^h + \bar{x} = \sum_{i=1}^N f_i(\bar{x}_i). \tag{69}$$

Here α_{hi} is the percentage of firm i that is owned by consumer h. Recall that shares are not traded in Becker's model.

Perhaps there is enough flexibility in (66) to permit more than one consumer to hold positive capital stock in steady state. We do not have space to analyze (66) rigorously.

What about concavity of (64)? At present useful sufficient conditions on β_h, u_h for concavity are hard to come by.

It is worthwhile to examine the class of U where there exists a Koopmans (Ref. 12) utility aggregate $V(\cdot, \cdot)$ such that

$$U(c_1, c_2, \ldots) = V(c_1, U(c_2, c_3, \ldots)). \tag{70}$$

It is easy to see from Koopmans (Ref. 12) that in the deterministic case this class is a general one that yields time-stationary optimal policy functions so that convergence of optimal paths to steady states may be discussed in an economy with time-stationary technology. The same is true in the stochastic case as well, provided that the utility function itself is nonrandom.

Examine the case in Section 2 where $\{r_t\}$ is i.i.d. and $N = 1$. In order to derive a generalized Bellman equation of the form

$$\begin{aligned} W(y_1) &= \max E_1 U(c_1, c_2, \ldots) \equiv \max E_1 V(c_1, U(c_2, c_3, \ldots)) \\ &= \max V(c_1, E_1 U) \\ &= \max V(c_1, E_1 W(y_2)), \end{aligned} \tag{71}$$

we must have $V(x, y)$ affine in y. But if $V(x, y)$ is continuous and concave in (x, y), then there must exist $A(x)$, $B(x)$ such that

$$U(c_1, c_2, \ldots) = A(c_1) + B(c_1)U(c_2, c_3, \ldots). \tag{72}$$

But (72) is of the form (64). Hence provided that the utility function is itself deterministic, utility functions of the form (64) appear to be the most general ones for which we can derive generalized Bellman equations in the uncertainty case. This observation may be important if analysts try to use these methods in the uncertainty case.

Magill (Ref. 15) has developed an asset-pricing model that builds on earlier work of (1) Bewley (Ref. 19) on general equilibrium theory with infinite-dimensional commodity spaces; (2) Brock and Magill (Ref. 26) on optimal growth in an economy driven by continuous-time stochastic processes; and (3) Lucas and Prescott (Ref. 27) on existence of a rational expectations equilibrium in an industry equilibrium context where the industry demand curve is random.

Magill's paper applies recent work on optimal stochastic control in continuous time by Bismut (Ref. 16) to develop an Arrow–Debreu model that possesses a recursive structure emanating from production's being driven by a vector of Ito processes. The paper focuses on the use of Bismut's duality characterization of optima in order to characterize equilibria.

Although Magill's work is still in progress, it is leading to a formalization of the notion of Keynes that the securities market generates enough information on the price of risk that each firm manager can use this information to solve a myopic maximization problem and still achieve global optimum. This type of question can be discussed by Magill's apparatus because his firms solve genuine optimization problems under adjustment costs, whereas the firms in our model (Sections 2–4) simply maximize profits under no costs of adjustment.

5. Continuous-Time Asset-Pricing Models

On a front independent of the work reviewed in Section 4, continuous-time asset-pricing models are being developed by Cox and Ross (Ref. 28), Cox, Ingersoll, and Ross (Ref. 10), and Hellwig (Ref. 29). These models are equilibrium descendants of the continuous-time "partial" equilibrium models like those of Merton (Ref. 1). We refer to Merton-type models (Ref. 1) as partial equilibrium rather than general equilibrium because these models do not trace the channels of interaction between the underlying sources of uncertainty in production and tastes as they feed into the equilibrium determination of asset prices. Hellwig's paper spells out this difference.

In the Merton-type models, which are dynamic generalizations of the Sharpe–Lintner capital asset–pricing model, claims to exogenously given random-dividend processes are competitively traded, and the model prices these claims. Not only does the type of model being reviewed here possess markets for claims to random dividends, but the random dividends themselves are determined endogenously through the interaction of conscious decisions by firms and consumers as well as through the physical material balance conditions of general equilibrium.

Hence the Brock (Ref. 7), Cox and Ross (Ref. 28), Cox, Ingersoll, and Ross (Ref. 10), and Lucas (Ref. 6) type of models being reviewed here can be integrated into the modern rational expectations macroeconomic framework built by Lucas (Ref. 30) Sargent (Ref. 31), and others to produce a model that is potentially capable of explaining observed empirical regularities in the interaction of stock prices and money Marsh (Ref. 32) is developing this more general model now.

In describing these models we shall use the discrete-time model developed as an organizing vehicle for discussing recent work in this area.

Hellwig pointed out that Merton's well-known model (Ref. 1) was partial equilibrium in the sense that his π_i processes were exogenously given and not related to the firm decision side of the model. Hence we need models where the π_i processes are endogenous.

Cox, Ingersoll, Ross (Ref. 10, hereafter referred to as CIR) examine a continuous-time version of this discrete-time model where the f_i are linear in x_i. They introduce markets for derivative claims such as options. The Nth activity is riskless and corresponds to borrowing and lending at a risk-free interest rate $r(t)$ that is itself endogenous. They develop a generalized Black–Scholes (Ref. 11) type of partial differential equation formula for the values of the derivative claims. CIR do this by positing the prices of the derivative claims to be a function of the underlying state of the economy, using necessary conditions like those of Merton's for an optimum portfolio from the consumers' side and using the boundary conditions that define the derivative claims. Ito's lemma is applied to the pricing function of the derivative claim in order to derive its corresponding partial differential equation.

Furthermore they solve an example for a closed-form solution for the path of the interest rate. The paper also contains a pricing formula for a European call option on a discount bond of maturity date s with exercise price K and expiration date T.

Cox and Ross (Ref. 28) have begun work on a continuous-time model analogous to the model of CIR and Lucas (Ref. 6), but I do not have enough of their new material to say any more about it.

6. Summary

This paper has summarized recent, mostly unpublished, work on asset-pricing models. We attempted to expose the underlying unity of approach that characterizes these models, in the hope of revealing the potential usefulness of rational expectations asset-pricing models under full information of the type discussed here.

It seems quite probable that in the future such models and their offspring will be used to attack questions of the following type.

1. What impact does an increase in the rate of inflation have on the quantity of "old" oil wells restarted under so-called turnkey contracts, and to what extent do welfare losses occur and how large are they? What effects will obtain on the claims to the wells and the price of the land on which they reside?

In the case of a turnkey operation the current start-up cost is written off the current year's income tax, but the revenues are received in the future. Clearly high-bracket investors should buy the well, write off the cost of starting it up, turn it into an income-producing asset, and sell it to a group of low-bracket investors. The seller realizes a capital gain (which is taxed at a lower rate in the process).

The models that have been reviewed here can be adapted to study the impact of an increase in the rate of inflation together with the concomitant increase in real progressivity of the income tax on the equilibrium price of, for example, such wells.

Stylized gains to trade of this type, which are induced by increased real progressivity of the income tax due to inflation, abound. What is the impact of an increase in the rate of inflation on the equilibrium price of such assets? The models reviewed here can be adapted to study such questions.

2. Any question that can be attacked with a general equilibrium model like that of Arrow–Debreu can be studied with these models with added advantages: (a) the prices of competitively traded claims to risky profits may be analyzed; (b) prices of derivative claims can be solved by using the techniques of Black and Scholes (Ref. 11), Merton (Ref. 1), and others; (c) impacts on the price of the service of systematic risk bearing may be studied and estimated by using the extra structure of these models.

Although nothing has been said in this survey about the important issues of imperfect information, we believe that the study of perfect information models of the type reviewed here will aid progress in this area.

There seems no doubt that the class of models reviewed here is rich enough to serve as a useful formulation for attacking outstanding issues in economic dynamics.

ACKNOWLEDGMENT

This research was partially supported by NSF Grant SOC 74–19692 to the University of Chicago. Responsibility for the contents is my own.

References

1. MERTON, R., An intertemporal capital asset pricing model, *Econometrica*, Vol. 41, pp. 867–887, 1973.
2. BROCK, W. A. and MIRMAN, L., Optimal economic growth and uncertainty: The discounted case, *Journal of Economic Theory*, Vol. 4, pp. 479–513, 1972.
3. BROCK, W. A. and MIRMAN, L., Optimal economic growth and uncertainty: The no discounting case, *International Economic Review*, Vol. 14, pp. 560–573, 1973.
4. RADNER, R., Optimal stationary consumption with stochastic production and resources, *Journal of Economic Theory*, Vol. 6, 1973.
5. JEANJEAN, P., Optimal development programs under uncertainty: The undiscounted case, *Journal of Economic Theory*, Vol. 5, pp. 66–92, 1974.
6. LUCAS, R. E., Jr., Asset prices in an exchange economy, *Econometrica*, Vol. 46, pp. 1427–1445, 1978.
7. BROCK, W., *An Integration of Stochastic Growth Theory and the Theory of Finance*, Department of Economics, University of Chicago, February 9, 1978.
8. ROSS, S., The arbitrage theory of capital asset pricing, *Journal of Economic Theory*, Vol. 13, pp. 341–359, 1976.

9. Prescott, E. and Mehra, R., *Recursive Competitive Equilibria and Capital Asset Pricing*, Carnegie Mellon University, Graduate School of Industrial Administration, December 1977.
10. Cox, J., Ingersoll, J. and Ross, S., *Notes on a Theory of the Term Structure of Interest Rates*, Stanford, Chicago, and Yale Universities (undated).
11. Black, F. and Scholes, M., The pricing of options and corporate liabilities, *Journal of Political Economy*, Vol. 81, pp. 637–659, 1973.
12. Koopmans, T., Stationary ordinal utility and impatience, *Econometrica*, Vol. 28, pp. 287–309, 1960.
13. Benveniste, L. M. and Scheinkman, J. A., *Duality Theory for Dynamic Optimization Models of Economics: The Continuous Time Case*, University of Chicago, revised August 1977.
14. Becker, R., Department of Economics, University of Rochester, Ph.D. thesis, 1978.
15. Magill, M., *Equilibrium and Investment Under Uncertainty*. Department of Economics, Northwestern University, 1978.
16. Bismut, J., Conjugate convex functions in optimal stochastic control, *Journal of Mathematical Analysis and Applications* Vol. 44, pp. 384–404, 1973.
17. Brock, W. A. and Majumdar, M., Global asymptotic stability results for multi-sector models of optimal growth under uncertainty when future utilities are discounted. *Journal of Economic Theory*, Vol. 18, pp. 225–243, 1978.
18. Loeve, M. *Probability Theory*, D. Van Nostrand Co., Princeton, N.J., 1963.
19. Bewley, T., Existence of equilibria in economies with infinitely many commodities, *Journal of Economic Theory* Vol. 4, pp. 514–540, 1972.
20. Mirman, L. and Zilcha, I., Characterizing optimal policies in a one-sector model of economic growth under uncertainty, *Journal of Economic Theory*, Vol. 14, pp. 389–401, 1977.
21. Mirman, L. and Zilcha, I., Unbounded shadow prices for optimal stochastic growth models, *International Economic Review*, Vol. 17, pp. 121–132, 1976.
22. Mirman, L. and Zilcha, I., On optimal growth under uncertainty, *Journal of Economic Theory*, Vol. 11, pp. 329–339, 1975.
23. Brock, W. A. and Gale, D., Optimal growth under factor augmenting progress, *Journal of Economic Theory*, Vol. 1, No. 3, pp. 927–929, 1970.
24. McFadden, D., On the existence of optimal development programmes in infinite horizon economics, *Models of Economic Growth*. Edited by J. Mirrlees and N. Stern, Halsted Press, New York, 1973.
25. Scheinkman, J. A., *Notes on Asset Pricing*, Department of Economics, University of Chicago, December 1977.
26. Brock, W. A. and Magill, M. J. P., Stability methods in dynamic economics, *Econometrica* Vol. 47, pp. 843–868, 1979.
27. Lucas, R. and Prescott, E., Investment under uncertainty, *Econometrica*, Vol. 39, pp. 659–682, 1971.
28. Cox, J. and Ross, S., *Some Models of Capital Asset Pricing with Rational Anticipations*, Stanford and Yale Universities, September 1977.
29. Hellwig, M., *A Note on the Intertemporal Capital Asset Pricing Model*, Department of Economics, Princeton University (undated).
30. Lucas, R. E., Jr., An equilibrium model of the business cycle, *Journal of Political Economy*, Vol. 83, pp. 1113–1144, 1975.
31. Sargent, T., A classical macroeconometric model for the United States, *Journal of Political Economy*, Vol. 84, pp. 207–237, 1976.
32. Marsh, T., University of Chicago, Ph.D. thesis, 1978.

Supplementary References

Brock, W. A., Money and Growth: The case of long-run perfect foresight, *International Economic Review*, Vol. 15, pp. 750–777, 1974.

Calvo, G., On models of money and perfect foresight, *International Economic Review* (to appear).

Dreze, J., *Investment under Private Ownership: Optimality, Equilibrium, and Stability*, C.O.R.E. Discussion Paper #7224. Center for Operations Research and Econometrics, Université Catholique de Louvain, Belgium.

MacClean, L., Field, C. A. and Sutherland, W., *Optimal Growth with Borrowing: The Stochastic Models*, Department of Mathematics, Dalhousie University, Halifax, N.S., 1976.

Malinvaud, E., Capital accumulation and efficient allocation of resources, *Econometrica*, Vol. 21, pp. 253–267, 1953.

Rubenstein, M., The valuation of uncertain income streams and the pricing of options, *Bell Journal of Economics*, Vol. 7, pp. 407–425, 1976.

2

Alternative Long-Run Goals and the Theory of the Firm: Why Profit Maximization May Be a Better Assumption Than You Think

HAYNE E. LELAND

1. Introduction

For several decades, economists have subjected the assumption that firms maximize profits to a barrage of criticism. Firms maximize sales (Ref. 1). Firms maximize utility of profits and sales (Ref. 2). Firms maximize just about anything *but* profits.† And firms that have alternative goals, we are led to believe, do not exhibit profit-maximizing behavior. In this paper, we accept the fact that firms may have goals that differ from profit maximization, but we reject the conclusion that their behavior must differ from that of a profit-maximizing firm.

Most alternative models of firm behavior assume a static environment: firms maximize current sales, utility of sales and profits, or whatever. The static framework is not surprising, since traditional profit-maximizing models also assume a static environment. But we shall argue that there is a crucial difference between profit maximization and other goals. For in the environments considered here, profit maximization over

† See Refs. 3 and 4. Some economists even deny that firms maximize anything. We shall assume that firms are rational and act as if they maximize *something*. Note that our analysis need not be limited to private ownership economies: Ward (Ref. 7) models the Yugoslav firm as maximizing profits per worker.

HAYNE E. LELAND • School of Business Administration, University of California, Berkeley, California.

many periods requires profit-maximizing behavior in the current period.† Such myopic behavior is not optimal in general for other objectives. Thus, Leland (Ref. 5) has shown that the revenue-maximizing firm does not maximize sales at each period, even though its long-run objective is to maximize sales. The answer to this paradox, we shall see, lies in the dynamics of firm growth.

In any economic system in which firms are liable for their debts, profits have a unique role: they provide the means by which firms can increase their capital stock. Therefore, profits may become an important indirect objective, if growth permits better achievement of the firm's goal in the future. This dynamic role of profits is perforce neglected by static models of the firm. In the case of long-run profit maximization, the static model is an accurate description because the direct and indirect goals coincide, and myopic behavior is optimal. But in the case of alternative long-run objectives, static models cannot be used to describe optimal firm behavior even in the current period.

In the following sections, we develop a model of firms with arbitrary long-run objectives. In doing so, we have two goals: first, to develop conditions under which optimal current behavior for these firms converges to profit maximization as the horizon becomes distant; second, to provide some simple tests that, when applied to a firm with any long-run objective, will indicate whether profit maximization is the optimal current policy. As examples of the general theory, we consider in some detail the cases of long-run utility of maximization of sales and profit and maximization of profit per worker.

We draw two important conclusions: (1) There are a large number of alternative goals whose optimal current policy requires profit maximization. Thus, profit maximization is a considerably more robust assumption than many believe. (2) Even if profit maximization is not the optimal current policy, current policies are likely to be closer (in a well-defined sense) to profit maximization policies than static models would indicate.

2. Formulation

Key elements of this theory include decision variables, a capital stock variable, and an objective functional which is the integral of some objective function.

† Following Jorgenson (Ref. 2), the presumed environment is one in which only current decision variables (e.g., labor hiring) and stock variables (e.g., "capital") directly affect current profits. Thus we rule out, for example, running at a loss in order to bankrupt a competitor: in that situation, past decisions (e.g., a decision that forced out a competitor) affect current profits directly, rather than indirectly through changes in current stock variables.

The decision variables, whose levels we denote by the n-dimensional vector $L(t)$, include all the variables that the firm can control from $t = 0$ to the terminal period, $t = T$. Typically, one thinks of labor or material inputs which are variable in the short run. More arcane variables, however, such as number of assistants directly responsible to the president, could also be included. For each period t, we assume that the choice levels $L(t)$ must belong to some feasible subset of R^n, denoted $\Lambda(t)$. This set, determined by physical or institutional constraints (if any), may be a function of current capital stock. An example of a constraint set is the set of labor input values at time t that exceed some minimum profitability as a percentage of capital stock—the "minimum profit constraint" of Baumol's saies-maximizing firm.

The capital stock variable is assumed to be a scalar, whose level is denoted by $K(t)$ and whose initial value is given: $K(0) = K_0$. (A subsequent section generalizes the results to the case in which there are multiple malleable capital goods.) Following most dynamic formulations, we assume the rate of change of capital stock $\dot{K}(t)$ is a function of net investment. Net investment, in turn, is assumed to be a function of current profits and capital stock. This function may shift with time. But for any level of capital stock and any time period, we presume that current net investment increases with current profit. This formulation describes a self-financing firm that reinvests all or a positive fraction of its profits. A more general approach would allow investment to be a decision variable, with the present value of investment limited by the present value of profits less dividends from $t = 0$ to $t = T$. In Appendix B, we show that many of the results we derive in the simpler, self-finance environment can be extended to a world with external finance.

The objective of the firm is assumed to be the integral of an arbitrary function whose arguments are the levels of capital stock, the decision variables, and time.† We represent such an objective by

$$\int_0^T F[K(t), L(t), t]\, dt.$$

Many objectives are a special case of this form; for example, a firm

† The integral criterion seems a natural extension of static goals to long-run objectives. It is the appropriate extension of the profit criterion, where the long-run objective is maximizing present value of profits or cash flow (see Jorgenson, Ref. 2, especially pp. 144–145). If managers' salaries are linearly related to some function of the firms' decision and stock variables, the maximization of present value of salaries can also be represented by maximization of an integral. But we do require that F, the current "felicity," depend only on *current* decision and stock variables, and not all preference rankings over functions through time can be represented by an integral; see the optimal growth literature on criterion functionals. Our analysis can readily be extended to include, in addition to the integral, a value function over terminal capital stock.

maximizing present value of utility of profits and sales would have

$$\int_0^T F[K(t), L(t), t]\, dt = \int_0^T e^{-rt} U\{P[K(t), L(t), t], S[K(t), L(t), t]\}\, dt.$$

No restrictions are placed on the form of market structure, the way in which the future is discounted (if at all), or the underlying technology of the firm. Any form is compatible with some function F.

Our formulation is summarized by assuming that the firm will

$$\text{Maximize} \int_0^T F[K(t), L(t), t]\, dt \tag{1}$$

subject to

$$\dot{K}(t) = I[P[K(t), L(t), t], K(t), t] \qquad \left(\frac{\partial I}{\partial P} > 0\right), \tag{2}$$

$$K(0) = K_0. \tag{3}$$

We further assume the following.

a. Given K_0, there exists an optimal policy function $L_T^*(t)$, for all t and T. The subscript denotes that, in general, the optimal policy function is a function of the horizon T.
b. Given $K(t)$, there exists a unique profit-maximizing policy $L^P(t) \in \Lambda(t)$. Note that $L^P(t)$ is not a function of T.
c. F and P are alternative goals in the sense that if, given $K(t)$, $L^P(t)$ lies in the interior of $\Lambda(t)$, then $L^P(t)$ does not also maximize F for $L(t) \in \Lambda(t)$.†

From the work of Pontryagin *et al.* (Ref. 6) we know that the optimal path $L_T^*(t)$ must satisfy, for all t,

$$\underset{L(t) \in \Lambda(t)}{\text{Maximum}}\ H(K, L, \lambda, t) = H(K, L^*, \lambda, t) \equiv H^*(K, \lambda, t), \tag{4}$$

when $\lambda(t)$ satisfies

$$\dot{\lambda} = \frac{-\partial H^*}{\partial K} \equiv -H_K^*, \tag{5}$$

$$\lambda(T) = 0, \tag{6}$$

and $K(t)$ satisfies (2) and (3), where H is the Hamiltonian

$$H(K, L, \lambda, t) = F(K, L, t) + \lambda I[P(K, L, t), K, t]. \tag{7}$$

† If L^P does not lie in the interior of $\Lambda(t)$, it may maximize F for $L \in \Lambda(t)$ even if F is an alternative goal.

Let $\lambda_T^*(t)$ denote the Lagrangian multiplier function that satisfies (5) and (6) when $L_T^*(t)$ is used.

3. Analysis of Optimal Policies

Our goal is not to solve explicitly for optimal paths $L_T^*(t)$ but to examine the firm's behavior at an arbitrary time t as the horizon becomes distant ($T - t \to \infty$). In particular, we seek conditions under which, as the horizon becomes distant, optimal current policy approaches profit-maximizing policy $[L_T^*(t) \to L^P(t)]$. We shall see that in many cases the convergence of current policies to profit maximization depends on the behavior of the objective functional as the horizon becomes distant. But to show this result requires several preliminary steps.

First, consider the following possible assumptions on the functions F and P and on the set Λ, for given $K(t)$ and t.

(i) F and P are continuous functions of L.
(i′) F and P are concave differentiable functions of L.
(i″) F and P are strictly concave, twice continuously differentiable functions of L.
(ii) $L^P(t) \in \mathring{\Lambda}(t)$, where $\mathring{\Lambda}(t)$ is the interior of $\Lambda(t)$.
(ii′) $L^P(t), L_T^*(t) \in \mathring{\Lambda}(t)$.
(iii) $\Lambda(t)$ is compact.
(iii′) $\Lambda(t)$ is convex and compact.

Note that (i″) → (i′) → (i), (ii′) → (ii), and (iii′) → (iii). Assumptions (i)–(i″) are regularly encountered in static theories of the firm. They are compatible with a wide range of alternative market or technological structures. Assumption (ii) states that the profit-maximizing policy always lies in the interior of the choice set: the firm always has the option of maximizing profits if it should wish. Assumption (iii) is equivalent to closedness and boundedness of the choice set, since $\Lambda(t)$ is in R^n. At the extreme, one could argue that the finiteness of the world's resources bounds the choice set. Assumption (iii′) is a straightforward restriction of (iii).

We may now prove the following theorem.

Theorem 3.1. Given (i) and (iii), then $\lim_{T-t\to\infty} \lambda_T^*(t) = \infty$ is a sufficient condition for $\lim_{T-t\to\infty} L_T^*(t) = L^P(t)$.

Proof. See Appendix A. □

Corollary 3.1. Given (i′), (ii), and (iii), then $\lim_{T-t\to\infty} \lambda_T^*(t) = \infty$ is a necessary and sufficient condition for $\lim_{T-t\to\infty} L_T^*(t) = L^P(t)$.

Proof. See Appendix A. □

Theorem 3.1 and its corollary indicate that the behavior of the marginal value of capital stock $\lambda_T^*(t)$ (measured in units of the objective) is critical to whether current profit maximization is optimal current policy. If $\lambda_T^*(t)$ becomes unboundedly large as the horizon recedes, profit maximization will be required. Under the assumptions of the corollary, the divergence of $\lambda_T^*(t)$ is also necessary if current optimal policies are to approach profit-maximizing policies.

The conditions under which the theorem and the corollary hold are satisfied by most dynamic economic models. For example, the model of the revenue-maximizing firm studied by Leland (Ref. 5) satisfies (i″), (ii), and (iii′). When the profitability constraint is not binding, (ii′) is satisfied as well.

The intuitive explanation of Theorem 3.1 and its corollary is straightforward. If the marginal value of capital stock becomes large, it will be worth sacrificing current performance (measured by the current level of F) in order to build up capital stock. But capital stock is built more rapidly by increasing profits. As the marginal value of capital stock becomes unboundedly large, it will be optimal to increase capital as rapidly as possible, leading the firm to maximize profits.

In many models, we can show that $\lambda_T^*(t)$ is an increasing function of T.† Even if it approaches a finite limit, we can still show that under certain conditions *optimal current policy moves closer to profit maximization as the horizon recedes*, when there is a single control variable L at each t. Let

$$D(t, T) = [L_T^*(t) - L^P(t)]^2,$$

where $D(t, T)$ is a distance function monotonically related to the Euclidean distance between $L_T^*(t)$ and $L^P(t)$.

Theorem 3.2. Given (i″), (ii′), and (iii′), then

$$\frac{dD(t, T)}{d\lambda_T^*(t)} < 0.$$

Proof. See Appendix A. □

If $\lambda_T^*(t)$ is an increasing function of time horizon T for given t, and if the conditions of Theorem 3.2 are satisfied, then optimal current policies

† If, for example, $F(K, L, t) = e^{-\sigma t}F^*(K, L)$ and $I[P(K, L, t), K, t] = I^*(K, L)$, then $\lambda_T^*(t)$ is a function of $T - t$ only. If along the optimal path $H_K^* > 0$, then (5) and (6) imply that $\lambda_T^*(t)$ is an increasing function of $T - t$.

will move toward the profit-maximizing policies—even if profit maximization itself is not reached in the limit. Theorem 3.2 elucidates the contention that static models in general yield incorrect current policies for firms that have alternative long-run objectives.

4. Extension to Multiple Capital Assets

Assume now that capital stock $K(t)$ is malleable and may be assigned to m sectors, with

$$\tilde{K}_i(t) = \alpha_i(t)K(t), \qquad i = 1, \ldots, m-1,$$

$$\tilde{K}_m(t) = \left[1 - \sum_{i=1}^{m-1} \alpha_i(t)\right] K(t).$$

For all t, $\alpha \in A$, where A is the compact convex set,

$$A = \{\alpha \in R^{m-1} \mid \alpha \geq 0, 1'\alpha \leq 1\}.$$

Let

$$P(\tilde{K}, L, t) = \bar{P}(K, \bar{L}, t), \qquad F(\tilde{K}, L, t) = \bar{F}(K, \bar{L}, t),$$

with $\bar{L} = (L, \alpha)$ and $\bar{\Lambda}(t) = \{L, \alpha \mid L \in \Lambda(t), \alpha \in A\}$. Theorem 3.1 and its corollary now apply, with $\bar{P}$, $\bar{F}$, $\bar{L}^*$, $\bar{L}^P$, and $\bar{\Lambda}(t)$ replacing P, F, L^*, L^P, and $\Lambda(t)$, respectively.

If capital is not malleable between the m sectors, the problem is more difficult and is not treated here.

5. Profit Maximization and Boundedness of the Criterion Integral

Theorem 3.1 and its corollary relate the behavior of the Lagrangian multiplier function $\lambda_T^*(t)$ to the behavior of the optimal limiting current policy. But unless H_K^* is bounded away from zero for all t as $T \to \infty$, it is difficult to determine the behavior of $\lambda_T^*(t)$ without first solving the optimization problem.

We shall therefore work toward characterizing the divergence of the current multiplier $\lambda_T^*(t)$ in terms of divergence of the criterion integral as $T \to \infty$. Consider the function

$$J[K(t), T] = \max \int_t^T F(K, L, \tau)\, d\tau$$

subject to

$$\dot{K} = I(K, L, \tau), \qquad K(t) \text{ given},$$

where K and L are vectors of arbitrary dimensions, and the $\Lambda(\tau)$ are the feasible control sets, conditional on $K(\tau)$, for $\tau \in [t, T]$. No restrictions are placed on the function I. Define

$$H(K, L, \lambda, \tau) = F(K, L, \tau) + \lambda(\tau)\, I(K, L, \tau),$$

$$H^*(K, \lambda, \tau) = \max_{L(\tau) \in \Lambda(\tau)} H(K, L, \lambda, \tau).$$

Let $\lambda_T^*(\tau)$ be the vector of functions satisfying

$$\dot{\lambda}_T^*(\tau) = -H_K^*, \qquad \lambda_T^*(T) = 0.$$

Theorem 5.1. For all $T < \infty$ and $t \leq T$, assume that

(i) $J[K(t), T]$ exists for all $K(t)$;
(ii) $H^*(0, \lambda_T^*, \tau) = 0$ for all $\tau \in [t, T]$;†
(iii) $H^*(K, \lambda_T^*, \tau)$ is $\left\{\begin{matrix}\text{concave}\\ \text{linear homogeneous}\\ \text{convex}\end{matrix}\right\}$ in K for all $\tau \in [t, T]$.

Then

$$J[K(t), T] \left\{\begin{matrix}\geq\\ =\\ \leq\end{matrix}\right\} \lambda_T^*(t) K(t) \quad \text{for all } t \leq T.$$

Proof. See Appendix A. □

Corollary 5.1. In addition to (i) of Theorem 5.1, assume that

(ii′) $H(0, 0, \lambda_T^*, \tau) = 0$, $\tau \in [t, T]$;
(iii′) $H(K, L, \lambda_T^*, \tau)$ is $\left\{\begin{matrix}\text{concave}\\ \text{linear homogeneous}\end{matrix}\right\}$ in K and L for all $\tau \in [t, T]$;
(iv) $L_T^*(\tau) \in \mathring{\Lambda}(\tau)$, $\tau \in [t, T]$, where $\mathring{\Lambda}(\tau)$ is the interior of $\Lambda(\tau)$.

Then

$$J[K(t), T] \left\{\begin{matrix}\geq\\ =\end{matrix}\right\} \lambda_T^*(t) K(t) \quad \text{for all } t \leq T.‡$$

† In fact, the limiting results of Proposition 5.1 can be shown to hold with the weaker assumption

$$\lim_{T-t\to\infty} \int_t^T H^*(0, \lambda_T^*, \tau)\, d\tau < \infty.$$

But assumption (ii) somewhat simplifies the analysis and will in fact be satisfied by most economic models, including those analyzed in Section 7. This follows from the fact that if $K = 0$, the constraint of nonnegative capital will require nonnegative profits, which in turn will require zero labor hiring as well as zero capital. This implies zero profits, and since output will be zero and remain at zero, we conclude that the value of the program itself will be zero. Of course, linear homogeneity of H^* in K always implies (ii).

‡ We have omitted the case of H convex in K and L since concavity of H in L is a necessary condition along the optimal path.

Proof. See Appendix A. □

Theorem 5.1 and its corollary relate $\lambda_T^* K(t)$, the imputed value of current stocks, to $J[K(t), T]$, the maximal value of the program from t until T. The relationships depend on the homogeneity or concavity properties of H^* (H) in K (K and L). From the corollary, note that if H is linear homogeneous in K and L, then

$$J[K(t), T] = \lambda_T^*(t)K(t) = \left[\frac{\partial J}{\partial K(t)}\right]K(t).$$

That is, if H^* (H) is linear homogeneous in K (K and L), then the optimal value function $J[K(t), T]$ is linear homogeneous in $K(t)$ for all T. *If the optimal control is used,* the total value of the optimal program will be imputed to the initial capital stock. Hence Theorem 5.1 can be considered a dynamic Euler theorem.

Our interest in Theorem 5.1 is that it provides a new way of examining convergence to profit maximization. Theorem 3.1 establishes a link between the shadow price of capital stock $\lambda_T^*(t)$ and convergence of optimal policies to profit-maximizing policies. Theorem 5.1 now establishes a link between $\lambda_T^*(t)$ and the value of the optimal program from t to T. We trace the resulting linkages in the following proposition.

Proposition 5.1. Given the assumptions of the corollary to Theorem 3.1 and the assumptions of $\left\{\begin{smallmatrix}\text{Theorem 5.1}\\ \text{its corollary}\end{smallmatrix}\right\}$, then for given (scalar) $K(t)$

(i) $\lim_{T-t\to\infty} J[K(t), T] = \infty$ is a necessary condition for $\lim_{T-t\to\infty} L_T^*(t) = L^P(t)$ if $\left\{\begin{smallmatrix}H^*\\ H\end{smallmatrix}\right\}$ is concave in $\left\{\begin{smallmatrix}K\\ K \text{ and } L\end{smallmatrix}\right\}$;
(ii) it is a necessary and sufficient condition if $\left\{\begin{smallmatrix}H^*\\ H\end{smallmatrix}\right\}$ is linear homogeneous in $\left\{\begin{smallmatrix}K\\ K \text{ and } L\end{smallmatrix}\right\}$.

Proof. See Appendix A. □

Proposition 5.1 demonstrates the close relationship between divergence of the objective integral on one hand and convergence to profit-maximizing policies on the other, as the terminal horizon becomes distant. Concavity or linear homogeneity of the Hamiltonian determines the exact nature of the relationship. This proposition is a generalization of Leland's Theorem IV (Ref. 5), which proved that divergence of the integral of sales was a necessary and sufficient condition for a sales-maximizing firm to use profit-maximizing policies. Constant prices through time, constant returns to scale, and a constant rate of time preference as well as a single control variable, were assumed in that study. The present proposition shows that only the second of those assumptions is required for the result, though prices must be invariant to output at each moment in time. If prices are not invariant to output but the sales function is strictly concave, then divergence remains a necessary though not sufficient condition.

6. Some Simple Tests for Convergence to Profit Maximization

The goal of this study is to discover whether profit maximization is optimal limiting behavior. We have demonstrated connections between the convergence of $L_T^*(t)$ to $L^P(t)$ and the limiting behavior of $J[K(t), T]$ as $T - t \to \infty$. But in general we cannot examine the behavior of J without computing $L_T^*(\tau)$ for increasing T, a tiresome procedure equivalent to a direct examination of whether $L_T^*(t)$ converges to $L^P(t)$.

In many cases, however, there exist readily computable functions that bound J above and below for all T. The limits of these functions therefore bound the limit of J as $T - t \to \infty$.

Define

$$F(K^F, L^F, \tau) = \max_{L \in \Lambda(\tau)} F(K^F, L, \tau), \tag{8}$$

where

$$\dot{K}^F = I[P(K^F, L^F, \tau), K^F, \tau],$$

$$K^F(t) = K(t),$$

and

$$P(K^P, L^P, \tau) = \max_{L \in \Lambda(\tau)} P(K^P, L, \tau), \tag{9}$$

where

$$\dot{K}^P = I[P(K^P, L^P, \tau), K^P, \tau],$$

$$K^P(t) = K(t).$$

Note that L^F, L^P, K^F, and K^P are not functions of T and are therefore relatively simple to compute for all τ. Now define

$$M[K(t), T] = \int_t^T F(K^P, L^P, \tau)\, d\tau, \tag{10}$$

$$N[K(t), T] = \int_t^T F(K^F, L^F, \tau)\, d\tau. \tag{11}$$

Because the paths $K^P(\tau)$, $L^P(\tau)$, and $K^F(\tau)$, $L^F(\tau)$ are feasible for all τ and T,

$$J[K(t), T] \geq M[K(t), T] \quad \text{and} \quad N[K(t), T] \qquad \text{for all } T. \tag{12}$$

Test A. Given the assumptions of Proposition 5.1, either

$$\lim_{T-t\to\infty} M[K(t), T] = \infty$$

or

$$\lim_{T-t\to\infty} N[K(t), T] = \infty$$

is a sufficient condition for

$$\lim_{T-t\to\infty} L_T^*(t) = L^P(t)$$

if H^* (H) is linear homogeneous in K (K and L).

Proof. See Appendix A. □

Now define

$$F(K^P, L^{FP}, \tau) = \max_{L\in\Lambda(\tau)} F(K^P, L, \tau), \qquad (13)$$

where K^P satisfies

$$\dot{K}^P = I[P(K^P, L^P, \tau), K^P, \tau], \qquad K^P(t) = K(t).$$

Define

$$S[K(t), T] = \int_t^T F(K^P, L^{FP}, \tau)\, d\tau. \qquad (14)$$

Test B. Given the assumptions of Proposition 5.1 and that $\max_{L\in\Lambda(\tau)} F(K, L, \tau)$ is a nondecreasing function of K for all τ,† then

$$\lim_{T-t\to\infty} S[K(t), T] = \infty$$

is a necessary condition for

$$\lim_{T-t\to\infty} L_T^*(t) = L^P(t)$$

when H^* (H) is concave or linear homogeneous in K (K and L).

Proof. See Appendix A. □

† This assumption implies that more capital will not hurt in achieving the objective F at any period in time. It is satisfied by all the models considered in Section 7. If this is not satisfied, then $\lambda \not\to \infty$.

Test A computes the profit-maximizing policies $L^P(\tau)$, the resulting capital stock path $K^P(\tau)$, and the value of the criterion integral $M[K(t), T]$ that results. But Test A is simple to run because, as T changes, we need not recompute $L^P(\tau)$; these policies, unlike $L_T^*(\tau)$, are invariant to T. Therefore it is much easier to examine the limiting behavior of M than the limiting behavior of J. (Similar comments apply to computing N.)

Test B also is simple to run. Using the capital stock path $K^P(\tau)$ computed in Test A, we now substitute $L^{FP}(\tau)$ for $L^P(\tau)$, where $L^{FP}(\tau)$ maximizes $F[K^P, L, \tau]$ for all τ. Note, as before, that $L^{FP}(\tau)$ is not dependent on the horizon T. The limiting behavior of S is therefore as simple to study as the limiting behavior of M or N.

7. Some Applications

7.1. A Firm Maximizing U(Profits, Sales)

Several authors have suggested a compromise between "pure" sales or profit maximization: a firm seeks to maximize $U(\Pi, S)$, where

$$\begin{aligned} &\text{Profits} \quad && \Pi(K, L) = pQ(K, L) - \omega L - rK, \\ &\text{Sales} \quad && S(K, L) = pQ(K, L). \end{aligned}$$

For simplicity, assume that p, ω, and r are constant through time and that U and Q are linear homogeneous in (Π, S) and (K, L), respectively. Extreme cases occur when $U(\Pi, S) = \Pi$ or S—the pure profit- or sales-maximizing models. An intermediate example is the Cobb–Douglas utility function $U = \Pi^\alpha S^{1-\alpha}$, $\alpha \in (0, 1)$. We assume that $\Lambda(t) = [0, \bar{L}(t)]$, where $\bar{L}(t)$ is some (large) bound, and that $\dot{K}$ is proportional to profit.

In a dynamic environment, the firm will

$$\text{Maximize} \int_0^T e^{-\sigma t} U(\Pi, S)\, dt$$

subject to

$$\dot{K} = m\Pi(K, L), \qquad m > 0, \qquad K(0) = K_0.$$

Given our assumptions,

$$H = e^{-\sigma t} U(\Pi, S) + \lambda m \Pi$$

is linear homogeneous in K and L, and

$$\int_0^T e^{-\sigma t} U(\Pi, S)\, dt = \int_0^T e^{-\sigma t} U^*(K, L)\, dt = \int_0^T e^{-\sigma t} K f(k)\, dt, \tag{15}$$

where

$$U[\Pi(K, L), S(K, L)] = U^*(K, L) = KU^*(1, L/K) = Kf(k),$$
$$k = K/L.$$

Applying Test A. $L^P(t)$ is chosen to maximize $\Pi(K^P, L)$. We assume $L^P(t) \in \mathring{\Lambda}(t)$. By homogeneity,

$$\Pi_L(K^P, L^P) = 0 \Rightarrow \Pi_L(K^P/L^P, 1) = 0 = \Pi_L^*(k^P),$$

where $k^P = K^P/L^P$. Note that k^P is independent of t. Using (15), we find that

$$M[K(0), T] = \int_0^T e^{-\sigma t} K^P f(k^P)\, dt = f(k^P) \int_0^T e^{-\sigma t} K^P\, dt,$$

which will diverge (converge) if

$$\dot{K}^P/K^P \geq (<)\, \sigma.$$

But

$$\dot{K}^P = m\Pi(K^P, L^P) = mK^P\Pi^{**}(k^P),$$

where $\Pi^{**}(k^P) = \Pi(1, L^P/K^P)$. By Test A, $m\Pi^{**}(k^P) \geq \sigma$ is a sufficient condition for

$$\lim_{T\to\infty} J[K(0), T] = \infty,$$

which by Proposition 5.1 implies

$$\lim_{T\to\infty} L_T^*(0) = L^P(0).$$

Applying Test B. $L^{FP}(t)$ is chosen to maximize $U^*(K^P, L)$. By homogeneity,

$$U_L^*(K^P, L) = 0 \Rightarrow U_L^*(K^P/L^{FP}, 1) = 0 = U_L^{**}(k^F),$$

where $k^F = K^P/L^{FP}$ and is independent of time. Using (15), we find that

$$S[K(0), T] = \int_0^T e^{-\sigma t} K^P f(k^F)\, dt = f(k^F) \int_0^T e^{-\sigma t} K^P\, dt,$$

implying (as in Test A) that $S[K(0), T]$ diverges if $m\Pi^{**}(k^P) \geq \sigma$. By Test B, $m\Pi^{**}(k^P) \geq \sigma$ is a necessary test for convergence of $L_T^*(t)$ to $L^P(t)$.

Combining the results of Tests A and B, we see that $m\Pi^{**}(k^P) \geq \sigma$ is a necessary and sufficient condition for profit maximization to be limiting optimal policy. The intuitive reasoning behind this result is that $\max \dot{K}/K = m\Pi^{**}(k^P)$. Therefore, the criterion integral diverges if and only if $\max \dot{K}/K \geq \sigma$. Even if $m\Pi^{**}(k^P) < \sigma$, it can be shown that $\lambda_T^*(t)$ is an increasing function of T for all t and, by Theorem 3.2, that $L_T^*(t)$ moves closer to $L^P(t)$ with increasing $T - t$, if $L_T^*(t)$ is a scalar for all t.

Note that the question of convergence to profit maximization is independent of the functional form of U, as long as U is linear homogeneous and k^F exists.

In fact, the condition $m\Pi^{**}(k^P) \geq \sigma$ is necessary and sufficient for any objective function that is stationary, linear homogeneous, and increasing in K and L, including sales maximization, when $\dot{K} = m\Pi$, and Π is stationary, linear homogeneous, and increasing in K and L. The *speed* at which $L_T^*(t) \to L^P(t)$ will, of course, depend on the objective function.

7.2. The Yugoslav Firm

Ward (Ref. 7) assumes that management by workers will lead the firm to maximize profit per worker, or, in a dynamic context, to

$$\text{Maximize} \int_0^T e^{-\sigma t}(\Pi/L)\, dt$$

subject to

$$\dot{K} = m\Pi = m[pQ(K, L) = \omega L - rK], \qquad K(0) = K_0.$$

If $Q(K, L)$ is linear homogeneous and prices are invariant to decision variables, (Π/L) will depend only on k and not on the absolute level of K. Therefore, the marginal value of capital stock, $\lambda_T^*(t)$, will be zero for all t and T, and myopic behavior will be optimal at all times: the static model is a complete description of dynamic behavior.

If $Q(K, L)$ exhibits increasing returns to scale, then capital stock will have a positive imputed value $\lambda_T^*(t)$, implying that *optimal behavior is closer to profit maximization than implied by a static model*, if L is a scalar. Whether profit maximization is optimal behavior in the limit depends on further specifications of the model, which we shall not introduce here.

8. Conclusion

We have shown that static models of firms with alternative goals neglect the dynamic role of profit and yield erroneous conclusions about current behavior of such firms. To the extent that growth enhances future achievement of a firm's arbitrary goal, profit becomes an indirect goal. This fact led us to examine the circumstances in which alternative long-run goals are best achieved by current profit maximization. For a large class of models, including the revenue-maximizing firm, the firm maximizing utility depending on profits and sales, and the workers' management firm, we have shown that divergence of the criterion integral as $T \to \infty$ is a necessary and in some cases a sufficient condition for current profit maximization to be optimal limiting policy. Even if the criterion integral does not diverge, we may still find current policies moving closer to profit maximization as the horizon becomes more distant.

One might be tempted to conclude, "Aha! Current optimal policy is never profit maximization, because this implies that there is no optimal policy for the infinite horizon." This conclusion should be resisted. First, it is not clear that we must require the existence of an optimal policy for the infinite horizon. Only inveterate optimists believe that the world will never end. If T is finite but very large, we may find that profit maximization is an excellent approximation of current policies. Second, there is no reason to limit our notion of optimality to the integral criterion. More general criteria, such as von Weiszacker's "overtaking criterion," are perfectly consistent with profit maximization as optimal limiting policy even when a criterion integral does not converge.

In sum, these results indicate that in many situations, including firms operating in noncapitalistic systems, profit maximization is consistent with the achievement of alternative long-run goals. To the extent that these situations describe the environment, *the simple assumption that firms seek alternative objectives does not warrant rejection of profit maximization as a description of current behavior.* And even in situations in which optimal limiting behavior is not profit maximization, optimal current policies are often closer to profit maximization than static models indicate.

ACKNOWLEDGMENT

This work was supported by National Science Foundation Grant GS-2874-A1 at the Institute for Mathematical Studies in the Social Sciences at Stanford University. The author wishes to thank Professors Takeshi Amemiya, Mukul Majumdar, and Scott Pearson for the generous time and advice they gave.

Appendix A

Proof of Theorem 3.1. For $\lambda > 0$, the $L^* \in \Lambda(t)$ that maximizes

$$\beta F(K, L, t) + I[P(K, L, t), K, t], \qquad \beta \equiv 1/\lambda, \tag{A1}$$

also maximizes $H(K, L, \lambda, t)$. Note that $\lambda \to \infty \Leftrightarrow \beta \to 0$, when $\lambda > 0$.†

Assume now that $L^* \not\to L^P$ as $\beta \to 0$. Then there exists a sequence $\beta_n \to 0$ and an associated sequence L_n^* that maximizes (A1) for each β_n, such that for some $\delta > 0$,

$$L_n^* \notin N_\delta(L^P) \quad \text{for all } n, \tag{A2}$$

where $N_\delta(L^P)$ is the open ball of radius δ about L^P.

By Assumptions (iii) and (i) in Section 3, F will be bounded on $\Lambda(t)$ and

$$\beta_n F(K, L_n^*, t) \to 0. \tag{A3}$$

Now define

$$\Lambda^*(t) \equiv \Lambda(t) - [N_\delta(L^P) \cap \Lambda(t)],$$

which is also compact. Again using (i), there will exist an $L^D \in \Lambda^*(t)$ such that profit $P(K, L, t)$ is maximized, $L \in \Lambda^*(t)$, at $L = L^D$. But

$$P(K, L^D, t) < P(K, L^P, t) \tag{A4}$$

by the assumption that L^P is the profit-maximizing policy in $\Lambda(t)$.‡

We now show that (A2) leads to a contradiction. The definition of L_n^* requires for all n that

$$\begin{aligned} &\beta_n F(L_n^*) + I[P(L_n^*)] \geq \beta_n F(L^P) + I[P(L^P)] \\ \text{or} \quad &I[P(L_n^*)] - I[P(L^P)] \geq \beta_n[F(L^P) - F(L_n^*)]. \end{aligned} \tag{A5}$$

Since $\partial I/\partial P > 0$, and $P(L_n^*) \leq P(L^D) < P(L^P)$, the left-hand side will be negative and bounded away from zero. But using (A3), the right-hand side of (A5) approaches zero. This leads to a contradiction of (A5) and therefore of (A2). □

Proof of Corollary 3.1. By Assumptions (i′) and (ii) in Section 3, L^P is a global maximum of P, $L \in \Lambda(t)$, and $P_L(L^P) = 0$. Now

$$H_L(L^P) = \beta F_L(L^P) + \frac{\partial I}{\partial P} P_L(L^P) = \beta F_L(L^P).$$

† When $\lambda < 0$, we would have to maximize the negative of (A1) in order to maximize H. Then $\lambda \to -\infty$ will imply that $L^* \to L^{MP}$, where L^{MP} is the $L \in \Lambda(t)$ that *minimizes* profit. This would be a consequence of capital growth's hindering rather than helping future achievement of the firm's objective.

‡ If L^P were a set of points rather than a singleton, the analysis could still be carried through, with convergence to L^P interpreted as convergence to some member of the maximizing set.

If $L_T^* = L^P$, then (ii) implies that $H_L(L^P) = 0$ or $F_L(L^P) = 0$. Therefore, by (i′) and (ii), L^P is a global maximum of F as well as $P, L \in \Lambda(t)$, given K and t. But this contradicts assumption (c), that F and P are alternative goals. Therefore, $L_T^* \neq L^P$ for all $\beta \neq 0$, and $\beta \to 0$ (i.e., $\lambda_T^* \to \infty$) is a necessary condition for $L_T^* \to L^P$. □

Proof of Theorem 3.2. Differentiating $D(t, T)$ with respect to λ_T^* gives

$$\frac{dD}{d\lambda_T^*} = 2(L_T^* - L^P)\frac{dL_T^*}{d\lambda_T^*}. \tag{A6}$$

From (ii′), $H_L(L_T^*, \lambda_T^*) = 0$ (given K and t). This and (ii′) imply that L_T^* can be expressed as a function of λ_T^* by the implicit function theorem, with

$$\frac{dL^*}{d\lambda_T^*} = -[H_{LL}(L_T^*)]^{-1}H_{L\lambda}(L_T^*) = -[H_{LL}(L_T^*)]^{-1}P_L(L_T^*)\left(\frac{\partial I}{\partial P}\right). \tag{A7}$$

Assumption (i″) implies that H_{LL} is negative. Using the mean value theorem, we find that

$$P_L(L_T^*) - P_L(L^P) = P_{LL}(\theta)[L_T^* - L^P], \tag{A8}$$

where $\theta \equiv \mu L_T^* + (1 - \mu)L^P$, $\mu \in (0, 1)$. By convexity, $\theta \in \Lambda(t)$. By (i″), $P_{LL}(\theta)$ is negative. By (ii′),

$$P_L(L^P) = 0. \tag{A9}$$

Substituting from (A7), (A8), and (A9) into (A6) gives

$$\frac{dD}{d\lambda_T^*} = -2H_{LL}^{-1}(L_T^*)P_{LL}(\theta)(L_T^* - L^P)^2\left(\frac{\partial I}{\partial P}\right) < 0, \tag{A10}$$

since H_{LL}^{-1} and $P_{LL} < 0$ and $\partial I/\partial P > 0$.† □

Proof of Theorem 5.1. By (4) and the definition of H,

$$J[K(t), T] = \int_t^T [H^*(K^*, \lambda_T^*, \tau) - \lambda_T^*\dot{K}\, d\tau, \tag{A11}$$

where K^* is the $K(\tau)$ associated with the optimal path. But

$$H^*(K^*, \lambda_T^*, \tau) \left\{\begin{matrix}\geq\\=\\\leq\end{matrix}\right\} H^*(0, \lambda_T^*, \tau) + [H_K^*(K^*, \lambda_T^*, \tau)]K^* \tag{A12}$$

because $H^*(K, \lambda_T^*, \tau)$ is $\left\{\begin{matrix}\text{concave}\\\text{linear homogeneous}\\\text{convex}\end{matrix}\right\}$ in K.

† The extension of this result to situations with an n-dimensional control vector would require that the product of two negative definite matrices be positive definite. While true in one dimension, this is not generally true for larger dimensions. The author thanks Professors Takeshi Amemiya and Michael Hurd for discussions on this point. It can be shown, however, that the profit associated with L_T^* increases with λ_T^*.

Noting that $H^*(0, \lambda_T^*, \tau) = 0$ by assumption, and substituting from (A12) into (A11) gives

$$J[K(t), T] \left\{\begin{smallmatrix}\geq\\=\\\leq\end{smallmatrix}\right\} \int_t^T \{[H_K^*(K^*, \lambda_T^*, \tau)]K^* - \lambda_T^* \dot{K}^*\}\, d\tau \tag{A13}$$

$$= \int_t^T -[\dot{\lambda}_T^* K^* + \lambda_T^* \dot{K}^*]\, d\tau, \qquad \text{using (5)},$$

$$= \lambda_T^*(t) K(t), \qquad \text{using (6)},$$

because H^* is $\left\{\begin{smallmatrix}\text{concave}\\\text{linear homogeneous}\\\text{convex}\end{smallmatrix}\right\}$ in K. □

Proof of Corollary 5.1.

$$H(K^*, L^*, \lambda_T^*, \tau)\left\{\begin{smallmatrix}\geq\\=\end{smallmatrix}\right\}H(0, 0, \lambda_T^*, \tau) + H_K^* K^* + H_L^* L^*$$

$$= H_K^* K^*, \qquad \text{using (ii') and (iv) in Section 5,}$$

since $L^* \in \mathring{\Lambda}$ implies $H_L^* = 0$ for all τ. Therefore

$$J[K(t), T]\left\{\begin{smallmatrix}\geq\\=\end{smallmatrix}\right\} \int_t^T [H_K^* K^* - \lambda_T^* \dot{K}^*]\, d\tau$$

$$= \int_t^T -[\dot{\lambda}_T^* K^* + \lambda_T^* \dot{K}^*]\, d\tau, \qquad \text{using (5)},$$

$$= \lambda_T^*(t) K(t), \qquad \text{using (6)},$$

because $H(K, L, \lambda_T^*, \tau)$ is $\left\{\begin{smallmatrix}\text{concave}\\\text{linear homogeneous}\end{smallmatrix}\right\}$ in K and L. □

Proof of Proposition 5.1. (i) By Theorem 5.1 or its corollary, $J \to \infty$ is a necessary condition for $\lambda_T^*(t) \to \infty$, if H^* (H) is concave in K (K and L). By the corollary to Theorem 3.1, $\lambda_T^* \to \infty$ is necessary and sufficient for $L_T^*(t) \to L^P(t)$. The condition $J \to \infty$ is therefore necessary for $L_T^*(t) \to L^P(t)$.

(ii) By Theorem 5.1 linear homogeneity implies that $J \to \infty$ is a necessary and sufficient condition for $\lambda_T^*(t) \to \infty$, which in turn implies that $J \to \infty$ is necessary and sufficient for $L_T^*(t) \to L^P(t)$. □

Proof of Test A. By (12), M or $N \to \infty$ implies that $J \to \infty$ as $T - t \to \infty$. But $J \to \infty$ implies $L_T^*(t) \to L^P(t)$ when H^* (H) is linear homogeneous in K (K and L) by Proposition 5.1(ii). Therefore, given the assumption of the test, M or $N \to \infty$ implies that $L_T^*(t) \to L^P(t)$. □

Proof of Test B. The definition of L^{FP}, Assumption (a) of the test, and the fact that $K^P(\tau) \geq K(\tau)$ for any feasible control sequence implies

$$F(K^P, L^{FP}, \tau) \geq F(K, L, \tau)$$

for all feasible paths (K, L), including those associated with $L_T^*(\tau)$. Therefore, for all T,

$$S[K(t), T] = \int_t^T F(K^P, L^{FP}, \tau)\, d\tau \geq \int_t^T F(K_T^*, L_T^*, \tau)\, d\tau = J[K(t), T],$$

where K_T^* is the path of K associated with controls L_T^*. Thus $\lim_{T-t\to\infty} S[K(t), T] < \infty \Rightarrow \lim_{T-t\to\infty} J[K(t), T] < \infty$, and by Proposition 5.1(i) and (ii), $\lim_{T-t\to\infty} L_T^*(t) \neq L^P(t)$. The necessity of $\lim_{T-t\to\infty} S[K(t), T] = \infty$ follows directly. □

Appendix B

Consider an alternative formulation of our problem, where there is an investment variable $I(t)$ which may be chosen at each moment t, subject to the constraint that the present value of investment cannot exceed the present value of profits plus terminal salvage value of capital stock. The rate of change of capital, $K(t)$, is assumed to be an increasing function of $I(t)$. Our problem now can be formulated

$$\text{Maximize} \int_0^T F[K(t), L(t), t]\, dt$$

subject to

$$\dot{K} = f(I, K, t), \qquad \frac{\partial f}{\partial I} > 0, \tag{B1}$$

$$\int_0^T e^{-rt} I(t)\, dt \leq \int_0^T e^{-rt} P(K, L, t)\, dt + e^{-rT} L[K(T)], \tag{B2}$$

$$K(0) = K_0,$$

where r is the rate of interest. The Hamiltonian of this system is

$$H(K, L, I, \lambda, \mu, t) = F(K, L, t) + \lambda f(I, K, t) + \mu e^{-rt}[P(K, L, t) - I]. \tag{B3}$$

If interior solutions are assumed, then necessary conditions for an optimal path are

$$\frac{\partial H}{\partial L} = F_L + \mu e^{-rt} P_L = 0, \tag{B4}$$

$$\frac{\partial H}{\partial I} = \lambda\left[\frac{\partial f}{\partial I}\right] - \mu e^{-rt} = 0, \qquad \mu \geq 0, \tag{B5}$$

where, as before, $\lambda(t)$ along the optimal path gives the shadow price of capital. Substituting from (B5) into (B4) gives

$$F_L + \lambda\left[\frac{\partial f}{\partial I}\right] P_L = 0. \tag{B6}$$

Theorems 3.1 and 3.2 follow directly, when $\lambda[\partial f/\partial I]$ replaces λ. Does $\lambda \to \infty$ imply that $\lambda[\partial f/\partial I] \to \infty$? If $[\partial f/\partial I]$ is bounded away from zero, or if $I \geq 0$ and the right-hand side of (b) is bounded as $T \to \infty$ when optimal policies are used, then the question can be answered affirmatively. Given these weak restrictions relating the behavior of optimal policies to the behavior of the imputed value of capital, Theorems 3.1 and 3.2 hold in the environment with external finance. Theorem 5.1 and Proposition 5.1 also hold, with the expanded Hamiltonian (B3) replacing the simpler Hamiltonian in the text.

While in many situations $L_T^*(t) \to L^P(t)$ for firms with alternative goals even in an environment with external finance, it does not follow from (B5) that the firm's *investment* $I_T^*(t)$ will necessarily converge to the investment that an identical profit-maximizing firm would choose. Thus we might be wiser to look at investment decisions rather than current input decisions in ascertaining whether firm's objectives differ from profit maximization.

References

1. Baumol, W., *Business Behavior, Value, and Growth*, Macmillan, New York, 1959.
2. Jorgenson, D., The theory of investment behavior, *Determinants of Investment Behavior*, Edited by R. Ferber, NBER, New York, pp. 129–155, 1967.
3. Marris, R. A., A model of "managerial" enterprise, *Quarterly Journal of Economics*, Vol. LXXVII, pp. 185–209, May 1963.
4. Williamson, O. E., *The Economics of Discretionary Behavior: Managerial Objectives in a Theory of the Firm*, Prentice-Hall, Englewood Cliffs, N.J., 1964.
5. Leland, H., The dynamics of a revenue maximizing firm, *International Economic Review*, Vol. 13, pp. 376–385, June 1972.
6. Pontryagin, L., Boltyanskii, V. G., Gamkrelidze, R. V. and Mischenko, E. F., *The Mathematical Theory of Optimal Processes*, John Wiley, New York, 1962.
7. Ward, B., Workers' management in Yugoslavia, *Journal of Political Economy*, Vol. LXV, pp. 373–386, 1957.

Supplementary References

Baumol, W., On the theory of the expansion of the firm, *American Economic Review*, Vol. LII, pp. 1078–1087, December 1962.

Penrose, E., *The Theory of the Growth of the Firm*, Wiley, New York, 1959.

Peston, M. H., On the sales maximization hypothesis, *Economica*, N.S., Vol. 26, pp. 128–136, 1959.

3

Mathematical Optimization and Economic Behavior

SHELDON S. L. CHANG

1. Introduction

Dynamic optimization adds a time dimension to the study of economic behavior. While maximization of profit and utility has been the microeconomic foundation for studying supply and demand, and a general equilibrium condition is the basis for an economic system, such idealized conditions do not exist in a changing world. There is a gap between the concepts *short run* and *long run*, namely, how does one shift from one position to the other?

While maximum principle and dynamic programming are still new concepts to many economists, they need not be. Capital goods are outstanding examples of state variables. Marginal long-range returns to capital are then the adjoint variables. Viewed in this light, Pontryagin's adjoint equations are no more than accounting equations for the marginal long-range return to capital.

In the subsequent paragraphs a general formulation that illustrates these points are given. We then proceed to three examples.

1.1. Problem Formulation

A class of economic problems can be formulated as follows:

$$\dot{x} = f(x, u, t), \tag{1}$$

SHELDON S. L. CHANG • Department of Electrical Sciences, State University of New York, Stony Brook, New York.

$$f_i(x, u, t) = \begin{cases} \alpha_i u_i - \delta_i x_i, & i = 1, 2, \ldots, k, \\ f_i(\mathbf{x}, \mathbf{u}, t), & i = k + 1, \ldots, n, \end{cases} \tag{2}$$

$$F(x, \mathbf{u}, t) = \sum_{i=1}^{i=k} u_i + c, \tag{3}$$

$$\int_0^T e^{-\rho t} g(x, c, t)\, dt + G(x(T))e^{-\rho T} = \max, \tag{4}$$

$$c \geq 0, \tag{5}$$

$$u_i \geq 0, \qquad i = 1, 2, \ldots, k, \tag{6}$$

where $x \in R^n$ and $u \in R^m$ are state and control variables respectively, and the components $x_1, x_2, \ldots, x_k$, $k \leq n$, represent k different kinds of capital goods. Equation (2) represents capital accumulation. In (2) and (3) $\mathbf{x}$ and $\mathbf{u}$ are subvectors consisting respectively of components x_i and u_i, with $k < i$, and c represents immediate consumption, or expenditure. Equation (3) is an accounting equation for resource allocation. In (4) the time horizon T may be infinity, in which case we impose the additional requirement that ρ be sufficiently large that the integral is bounded.

Pontryagin's maximum principle gives a necessary condition for optimum control u^*. Let $q(t) \in R^n$ and a scalar function $H(q, x, u, t)$ be defined as follows:

$$H(q, x, u, t) = g(x, c, t) + q(t)f(x, u, t), \tag{7}$$

$$\dot{q}(t) = -\frac{\partial H(q, x, u, t)}{\partial x} + \rho q(t), \tag{8}$$

$$q(T) = \left\{\frac{\partial G(x)}{\partial x}\right\}_{x = x(T)} \tag{9}$$

The maximum principle requires that

$$H(q^*, x^*, u^*, t) = \max_{u \in A} H(q^*, x^*, u, t), \tag{10}$$

where x^*, q^* are the solutions of (1), (8), and (9) with $u = u^*$.

In these equations x and f are regarded as column vectors, while q, $\partial H/\partial x$, and $\partial G/\partial x$ are row vectors. If $T = \infty$, then (9) is to be replaced by a steady-state optimality condition. In applying the maximum principle we regard c as a dependent variable defined by (3). Conditions (5) and (6) are regarded as defining A, the region of allowable u.

To obtain a necessary condition on u^*, we are dealing with only the optimal pair x^* and q^*, and the superscript * can be omitted. The

Hamiltonian function can be written as

$$H(q, x, u, t) = g(x, c, t) + \mathbf{qf}(\mathbf{x}, \mathbf{u}, t) + \sum_{i=1}^{i=k} q_i(\alpha_i u_i - \delta_i x_i). \tag{11}$$

The first k components of (8) give

$$\dot{q}_i = -g_i + (\delta_i + \rho)q_i, \qquad i = 1, 2, \ldots, k, \tag{12}$$

where g_i denotes $\partial g/\partial x_i$. From (3) we have

$$\frac{\partial g(x, c, t)}{\partial u_i} = \left(\frac{\partial g(x, c, t)}{\partial c}\right)\left(\frac{\partial c}{\partial u_i}\right) = -g_c. \tag{13}$$

Let M denote the maximizing set; then $i \in M$ means that

$$q_i\alpha_i = q_m \triangleq \max_{j \leq k} \{q_j\alpha_j\} \tag{14}$$

and $i \leq k$. To maximize H, set

$$u_i = 0 \qquad \text{for } i \notin M. \tag{15}$$

Equation (10) can be satisfied in three ways:

$$-g_c + q_m = 0; \tag{16}$$

$$-g_c + q_m > 0, \qquad c = 0, \qquad \sum_{i \in M} u_i = F(x, \mathbf{u}, t); \tag{17}$$

$$-g_c + q_m < 0, \qquad u_i = 0, \qquad c = F(x, \mathbf{u}, t). \tag{18}$$

The significance of q_i is seen from dynamic programming. Let $V(x, t)$ denote the return function

$$V(\xi, t) = \max_{u(\cdot)} \left\{\int_t^T e^{-\rho(\tau - t)} g(x, c, \tau)\, d\tau + G(x(T))e^{-\rho(T-t)}\right\}, \tag{19}$$

where the maximizing is over all allowable $u(\tau)$ with $t \leq \tau \leq T$ and the initial condition $x(t) = \xi$. Then

$$q_i(t) = \frac{\partial V(x, t)}{\partial x_i}, \tag{20}$$

which is the marginal increase in future return with each unit of x_i added. Equations (16)–(18) give a criterion for optimum distribution between investments and immediate expenditure c: The long-range marginal return from an investment is equal to the immediate marginal return from c.

2. Energy Policy (Refs. 1–3)

The model we adopt for energy policy studies is a neoclassical growth model with energy services included in the production function for general goods as a third factor of production. While energy has long been recognized as indispensable in a productive process, an implicit assumption in the two-factor production function is that energy itself can be produced in any required amount by capital and labor, and a separate accounting of energy services is not necessary. However, depletion of energy resources is a focal problem to be studied here, and such a simplifying assumption cannot be made.

Other modifications from prior optimum growth models are the introduction of depletion factors for the production of energy goods, and inclusion of energy consumption in the felicity function. The latter is necessary to account for the trade-off between direct consumption of energy goods and use of energy goods in production.

2.1. System Equations

The problem of energy resource constraint can be formulated as follows:

$$F(K_0, L_0, R, t) = \sum_{i=0}^{i=N} I_i + e(R_0) + Lc, \tag{21}$$

$$\dot{K}_i = I_i, \qquad i = 0, 1, 2, \ldots, N, \tag{22}$$

$$R = \sum_{i=0}^{i=N} R_i - Lr, \tag{23}$$

$$R_i = G_i(K_i, g(K_i)L_i, t)\phi(D_i), \tag{24}$$

$$\dot{D}_i = -R_i, \qquad i = 1, 2, \ldots, N, \tag{25}$$

$$\dot{L} = bL, \tag{26}$$

$$L = \sum_{i=0}^{i=N} L_i, \tag{27}$$

$$J = \int_0^\infty e^{-\rho t}(LU(c, r, b) - H(R_0))\, dt = \max. \tag{28}$$

It is assumed that there are N different types of energy resources R_i, $i = 1, 2, \ldots, N$. Each type R_i has total deposits (known or unknown) D_i, capital equipment K_i, and labor force L_i for its production. Note that L_i includes only laborers working directly on production of R_i. The equipment K_i is lumped together into a *general goods* item, which includes everything else except R_i, $i = 1, 2, \ldots, N$, and L_0, K_0 represent the labor and capital respectively for production of general goods. The energy resources R_i, $i = 1, 2, \ldots, N$, are assumed to be exchangeable in the production of general goods, and the total energy resource for such use is denoted by R.

Equation (21) means that the total production of general goods is channeled into net investment I_0 for the general goods productive capital K_0, investments I_i for the capitals K_i, expenditure $e(R_0)$ for exchanging energy resource R_0 from some external source, and consumption Lc. Equation (22) represents that capital of various types are accumulated from corresponding net investments. Equation (23) gives an accounting of R as the sum of various resources R_i (including R_0), with direct consumption of energy Lr deducted.

In (24) the term $G_i(K_i, g(K_i)L_i, t)$ represents the mining effort for resource R_i with capital K_i and labor L_i. The function $g(K_i)$ is an increasing function of K_i and represents the learning-by-doing factor (Refs. 4, 5). The term $\phi_i(D_i)$ decreases as D_i decreases. It represents decreasing returns as lower-grade mines are used when a resource R_i is being depleted. Equation (25) gives the relation between D_i and R_i.

To clarify the meanings of these terms, I give some typical resources as illustrations.

1. *Oil wells.* The techniques are well developed: $g(K_1) = 1$, and $\phi_1(D_1)$ represents decreasing returns as D_1 is reduced. At first only the gushers are used. Then the industry extends to deeper wells and marginal fields.

2. *Offshore drilling and shale oil.* New techniques are being developed, and $g(K_2)$ increases with more investment K_2, representing the improving technology. However, its deposit is finite and $\phi_2(D_2)$ is a decreasing function as D_2 decreases.

3. *Solar energy.* New techniques are being developed but are very far from commercially applicable on a large scale. The resource is unlimited, and $\phi_3(D_3) = 1$.

4. *Nuclear fusion.* A workable technique for peaceful use of nuclear fusion energy has not been developed. However, it is promising. The resource is virtually unlimited if developed, and $\phi_4(D_4) = 1$.

Equations (26) and (27) represent the growth of labor population and division of labor into general production and production of various types of energy resources.

In (28) the felicity function U now depends on three variables, per capita consumption c, per capita consumption of energy resources r, and birth rate b. The dependences of U on c and r are expected to be rising with diminishing return. The function U should be quite flat for a reasonable range of b, but it decreases if b is required to be either too large or too small. The term $H(R_0)$ represents noneconomic penalty for dependence on foreign energy resources. It is a sharply rising function of R_0. The integral J represents the total felicity for the entire population with some discount on future returns. The future discount rate ρ can be explained by the uncertainty of looking into the future and that we do not care about posteriority as much as we do the present, as well as any number of other reasons.

2.2. Optimization

In the system (22), the equations (25) and (26) are state equations; K_i $(i = 0, 1, 2, \ldots, N)$, D_k $(k = 1, 2, \ldots, N)$, and L are state variables; and the other equations are constraints that reduce the number of independent control variables. Without loss of generality, the following are taken as control: I_i $(i = 0, 1, 2, \ldots, N)$, L_i $(i = 1, 2, \ldots, N)$, R_0, r, and b. The variables L_0, R_i $(i = 1, 2, \ldots, N)$, R, and c are regarded as dependent functions of the state and control variables as expressed by (27), (24), (23), and (21), respectively.

Let q_i and p_i denote the adjunct variables for K_i and D_i respectively. Equations (22) and (21) are of the form (2) and (3). The variables K_i and q_i can be identified with $x_i u_i$, $i \leq k$. It is assumed that the various types of productive capital K_i are at their optimum values, and (14), (17) give

$$q_i = U_c, \qquad i = 0, 1, 2, \ldots, N. \tag{29}$$

From (8) and (10) the following equations are obtained:

$$-\frac{\dot{q}}{q} = \frac{d}{dt} \ln U_c = F_{K_0} - \rho, \tag{30}$$

$$F_{K_0} = (F_R - p_i/q)[G_{ik} + G_{iL} g'(K_i) L_i]\phi_i, \tag{31}$$

$$F_{L_0} = (F_R - p_i/q)[G_{iL}\phi_i g_i(K_i)], \tag{32}$$

$$F_R = e'(R_0) + H'(R_0), \tag{33}$$

$$U_r = U_c F_R, \tag{34}$$

$$\frac{\partial V(K, D, L, T)}{\partial K_i} = U_c, \quad i = 1, 2, \ldots, N. \tag{35}$$

2.3. Interpretations of Mathematical Results

The policy implications of the mathematical results will now be discussed.

1. *Free-market condition for energy resources.* In (34), U_r and U_c are marginal felicities of energy resources and consumption goods. The ratio

U_r/U_c represents the price that users are willing to pay for energy resources in terms of consumption goods, and F_R is the additional units of consumption goods that can be produced if an additional unit of energy resource is used in production. Equation (34) means that price of oil to consumer should equal the price of additional goods produced if the same amount of oil is used in production.

2. *Import duties on energy resources.* Equation (33) means that more than ordinary import duties should be applied to energy resources. Because price rises when demand increases and drops when supply increases, the marginal cost $e'(R_0)$ is higher than the average import price $e(R_0)/R_0$.

Because the price of imported and domestic oil must be equal in a free market, the difference $H'(R_0) + e'(R_0) - e(R_0)/R_0$ represents the import duty to be levied.

3. *Equal marginal utility of all investments.* Equation (35) shows that marginal utility of investment should be equal to the marginal felicity of consumption for all types of investments. However, because the utility function is difficult to calculate, this result has only conceptual value and is difficult to apply.

4. *Capital rental and interest rate.* Equation (30) gives the relation between capital rental F_{K_0} and interest rate. In an economy with rising income, U_c is expected to decrease due to increasing c. A unit of goods consumed today would yield a utility $U_c(t)$. The same goods with interest at Δt later will yield a utility

$$e^{i\Delta t}U_c(t + \Delta t)e^{-\rho\Delta t},$$

where i is the interest rate. Under rule of free substitution, this expression is equal to U_c:

$$e^{i\Delta t}U_c(t + \Delta t)e^{-\rho\Delta t} = U_c(t). \tag{36}$$

Solving i from (36) and letting $\Delta t \rightarrow 0$ gives

$$i = -\frac{d}{dt}\ln U_c + \rho. \tag{37}$$

Equations (37) and (30) give

$$i = F_{K_0}. \tag{38}$$

5. *Tax on production of depletable resources.* Under ideal conditions the wage of a worker is his marginal production for general goods F_{L_0};

$$w = F_{L_0}. \tag{39}$$

If we assume that workers for different products are exchangeable, then the same wage is paid to a worker who marginally produces $G_{iL}\phi_i g_i(K_i)$

units of ith resources. Therefore the price at which a manufacturer can offer the ith resources is

$$\frac{w}{G_{iL}\phi_i g_i(K_i)} = F_R - \frac{p_i}{q}. \tag{40}$$

According to (32), because F_R represents the market price, p_i/q is the tax that can be levied on the product.

6. *Investment incentive.* Equation (31) can be expressed as follows: marginal product of capital = net receipt from oil sales after tax × marginal oil production of capital. Let M_0 denote the marginal oil production of capital. From (40)

$$M_0 = [G_{ik} + G_{iL}g_i'(K_i)L_i]\phi_i. \tag{41}$$

In this expression oil is used in the sense of an unspecified energy resource. Suppose N firms are engaged in the new industry of exploration and production of shale oil:

$$K_i = \sum_\mu K_{\mu i}, \qquad L_i = \sum_\mu L_{\mu i}.$$

The production function for the μth firm is

$$Q_\mu = G_i(K_{\mu i}, g_i(K_i)L_{\mu i})\phi_i(D_i). \tag{42}$$

A firm making an investment *unilaterally* would obtain a marginal increase in production:

$$\frac{\partial Q_\mu}{\partial K_{\mu i}}[G_{iK} + G_{iL}g_i'(K_i)L_{\mu i}]\phi_i. \tag{43}$$

Comparing (41) and (43), we see that the difference in return to investment must be financed by the government in some form of investment incentive:

$$M_0 - \frac{\partial Q_\mu}{\partial K_{\mu i}} = G_{iL}g_i'(K_i)(L_i - L_{\mu i})\phi_i \tag{44}$$

$$\simeq G_{iL}g_i'(K_i)L_i\phi_i.$$

7. *Government-organized or -sponsored research and development of alternative energy resources.* An unproved or undeveloped energy resource can be represented by the condition that

$$F_{L_0} > F_R G_{iL}\phi_i g_i(K_i) \tag{45}$$

for all values of L_i. Then $L_i = 0$ and there is no marginal return to investment. However, $\partial V/\partial K_i \neq 0$, and can be quite large. Such investments are usually called long range research and development projects.

They have no prospect of immediate financial return. This gap can be filled by government-organized or -financed research.

Items 1, 3, and 4 are conditions that normally exist under a free or mixed economy. Items 2, 5, 6, and 7 constitute a well-balanced and coordinated long-range energy policy.

3. Consumption Dynamics

While the utility theory of consumption is well developed, an area little explored is the dynamics of consumption. What happens if there is a sudden change of income expectation? Does the consumer distribute his purchases in much the same way that he would if he had had the new income all along? When does he tend to save or to go into debt?

3.1. Mathematical Model

The consumer is assumed to maximize consciously or unconsciously his integrated utility function

$$J = \int_0^\infty e^{-\rho t} u(c, K)\, dt, \tag{46}$$

where c and K are vectors representing the various expendable and durable goods:

$$\dot{K}_i = I_i - d_i K_i, \qquad i = 1, 2, \ldots, n, \tag{47}$$

$$y = \sum_j c_j + \sum_i I_i, \tag{48}$$

$$I_i \geq 0, \qquad i = 1, 2, \ldots, n - 1, \tag{49}$$

$$c_j \geq 0, \qquad j = 1, 2, \ldots, m'. \tag{50}$$

In (47), K_n represents total money saved, I_n represents current savings, and d_n is negative with $-d_n$ being the bank interest rate. The other K_i represents stocks of various durable goods which deplete at rates d_i. Equation (48) is an accounting equation stating that the net income y is either spent on expendable or durable goods or saved.

3.2. Optimization

Equations (46)–(50) represent a special case of (1)–(6) with $k = n$, $I = u$, $K = x$, and $\alpha = 1$. Equations (11)–(20) are easily rewritten in terms of the corresponding economic variables. In (48), c_i with $i \neq 1$ and

I_j for all j are considered control variables. Maximizing H with c_i gives

$$U_{c_i} - U_{c_1} = 0, \qquad i = 1, 2, 3, \ldots, n.$$

Therefore we can omit the subscript i and write $U_{c_i} = U_c$. From (16),

$$q_m = U_c = U_{c_i}. \tag{51}$$

Noting that U_{K_i} is denoted U_i, we can rewrite (12) as

$$\dot{q}_i/q_i = -U_i/q_i + (d_i + \rho). \tag{52}$$

Let ψ denote $\dot{q}_m/q_m$. Then from (51) and (52)

$$\psi = \frac{d}{dt} \ln U_c, \tag{53}$$

$$U_i = (\rho + d_i - \psi) U_c, \qquad i \in M. \tag{54}$$

Equations (51) and (54) give optimum distribution of purchases among expendable and durable goods. Under equilibrium condition, $\psi = 0$, and the distribution reduces to a Pareto optimum.

If $T = \infty$, the transversality condition (9) is replaced by (51) and (54) with $\psi = 0$ at $T = \infty$.

Example 3.1. The function $U(c, K)$ is of logarithmic Cobb–Douglas form, $n = m = 1$, and y = constant:

$$U = (1 - \alpha) \log c + \alpha \log K. \tag{55}$$

Then $U_c = (1 - \alpha)/c$, and $U_K = \alpha/K$. The condition (17) cannot happen. The solution under (18) is trivial: $c = y$. Under (16), (52) gives

$$\dot{c} = \frac{\alpha c^2}{(1 - \alpha) K} - (d + \rho) c. \tag{56}$$

Equation (47) can be rewritten as

$$\dot{K} = y - c - dK. \tag{57}$$

The steady-state solution at $t = \infty$ is obtained by assuming $\dot{c} = \dot{K} = 0$:

$$c_0 = \frac{(1 - \alpha)(d + \rho) y}{d + \rho - \alpha\rho}, \tag{58}$$

$$K_0 = \frac{\alpha y}{d + \rho - \alpha\rho}. \tag{59}$$

The optimum trajectory satisfies (56) and (57) at all times and (58) and (59) at $t = \infty$. However, it will not do to use (58) and (59) as the initial condition and then integrate (56) and (57) along $-t$, because the

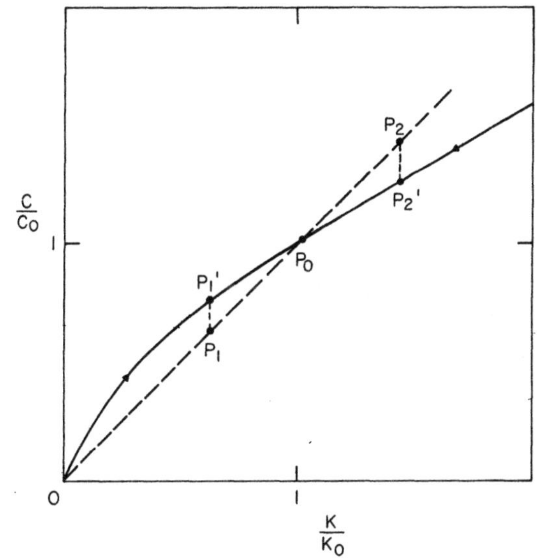

Figure 1. Optimum consumption as a function of investment.

right-hand sides of both equations vanish. However, it is readily shown that using $c_0 + \varepsilon_c$ and $K_0 + \varepsilon_K$ as initial values and integrating backwards gives an approximation to the optimal trajectory, which approaches the latter as $\varepsilon_c \to 0$ and $\varepsilon_K \to 0$. The result is shown in Fig. 1.

Suppose that the consumer is initially at equilibrium point P_0 and receives an unexpected salary raise. His c_0 and K_0 increase proportionately with operating point moving from P_0 to P_1. Since c can be changed at will, the new optimum operating point is now P_1'. The consumer increases his c immediately but not to its eventual optimum. He builds up his durable goods K so that he has equal satisfaction from both K and c.

Example 3.2. The function $U(c, K)$ satisfies (55), with $m = 2$, $d_1 > d_2$, and $y =$ constant.

Essentially the same procedure of finding the steady-state optimum and integrating backwards yields an optimum trajectory $c^*(t)$, $K_1^*(t)$, and $K_2^*(t)$. Plotting K_1/K_{10} and K_2/K_{20} gives the state diagram of Fig. 2, where K_{10} and K_{20} are $K_1^*(\infty)$ and $K_2^*(\infty)$ respectively.

Case 1. K_2 does not represent money savings. The optimum trajectories Γ_1 and Γ_2 divide the state plane into two regions. In the upper region K_1 is overinvested relative to K_2 and $I_1 = 0$. In the lower region K_2 is overinvested and $I_2 = 0$. Neither K_1 nor K_2 can make an instant finite change.

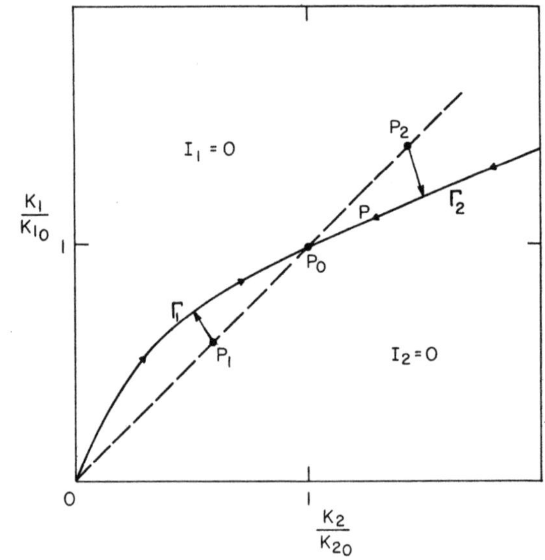

Figure 2. Optimum trajectory in the phase plane of durable goods.

Suppose that the consumer is initially at P. He uses his income for I_1 and c initially, until he reaches the Γ_2, then he follows Γ_2 to P_0.

Suppose that the consumer receives an unexpected salary raise and his operating point is now at P_1. He initially leaves K_2 as is, and uses his income for c and the less durable K_1. Eventually he increases both K_1 and K_2 after his operating point reaches Γ_1.

Case 2. K_2 represents money savings

$$K_2 = D_L + S, \tag{60}$$

where S is the actual money savings and D_L represents debt limit. A consumer at P_1 withdraws part of his savings or goes into debt to make an initial purchase of durable goods and gradually increases c, K_1, and K_2 along Γ_1.

4. Investment Behavior of Data-Processing Users (Ref. 6)

Investment in data processing is different from investments in machinery and plant; the difference is that the major and more permanent part of the data-processing investment is not so much the computing equipment but the application programs. It is increasingly so as years go by. The economic implications of this will be analyzed.

The mathematical model used in this study is as follows.

Production Function

$$Q = F(K, L, A, t), \tag{61}$$

where Q is the annual production of goods and services of a given firm, K and L are productive capital and labor employed by the firm, A is a measure of application programs being used in the productive process, t is the time variable, and F is the production function. The advance of new computer applications is represented by the dependence of F on A and t.

Figure 3 is a typical plot of Q/L versus K/L at a given time, with A as a parameter. The solid lines represent the situation at time t. There are $N(t)$ known application programs that can be used to increase the productivity Q/L. The curve Q/L goes up as A increases, and the upper limit of A is $N(t)$. The same plot at a later time t', $t' > t$, is represented by the broken lines. For given values of K/L and A, Q/L at t' is higher than Q/L at t. The increments represent general technological progress and improvements of existing application programs over the period $t' - t$. In the meantime the number of known applications also increases: $N(t') \geq N(t)$. The general rise of Q/L curve and increase of $N(t)$ are continuous processes of time.

The price p of the firm's product is a diminishing function of Q. Let G denote the product pF. Then G is a function of K, L, A, and t:

$$G(K, L, A, t) = pF(K, L, A, t).$$

In a more general sense G is the firm's output flow of goods and services in money terms.

Computing Facilities

The total amount of computing facilities is denoted by C. It consists of two parts C_1 and C_2:

$$C = C_1 + C_2, \tag{62}$$

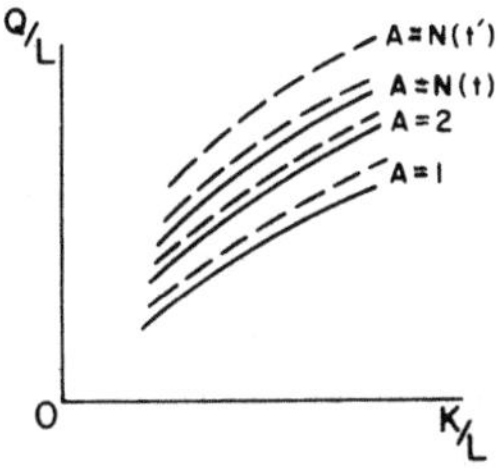

Figure 3. Production function with A as parameter.

where C_1 represents capital investment in computing facilities and C_2 is obtained under rental agreement. The accumulation of C_1 gives

$$\frac{dC_1}{dt} = I_C - hC_1. \tag{63}$$

The capital C_1 is treated as an entity separate from K and is not included in the latter, and hC_1 represents maintenance and depreciation of C_1.

Development of Application Programs

$$\frac{dA}{dt} = \Phi(C - \alpha(A), P - \beta(A)). \tag{64}$$

In (64), P represents the number of data-processing personnel (programmers), $\alpha(A)$ and $\beta(A)$ represent the computer facility and manpower required for operating and maintaining the application programs currently in use. The function Φ takes the form of a production function, with $C - \alpha(A)$ taking the place of K, and $P - \beta(A)$ taking the place of L.

The Gross Profit, Net Profit, and Investment

The gross profit E_g, net profit E_n, and accumulation of capital K are given as

$$E_g = G(K, L, A, t) - sL - wP - rC_2, \tag{65}$$

$$\dot{K} = I_k - \delta K, \tag{66}$$

$$E_n = E_g - T(E_g, I) - I, \tag{67}$$

$$I = fI_C + gI_k, \tag{68}$$

$$\frac{\partial T}{\partial E_g} > 0, \qquad \frac{\partial T}{\partial I} \leq 0, \tag{69}$$

where

I = money worth of investment,
T = business tax,
s = wage of labor,
w = wage of programmer,
r = rental for computing facilities C_2,
f = price for computing facilities,
g = price for other productive capital,
δ = maintenance and depreciation of other productive capital K.

Net Worth of the Firm

The net worth of the firm is the total of the discounted value of its future net earnings (Refs. 7, 8):

$$W(t) = \int_t^\infty \exp\{-\gamma(t' - t)\}E_n(t')\,dt', \tag{70}$$

where γ is a value larger than the interest rate to account for both the interest rate and uncertainty in future events.

A fundamental assumption of investment behavior is the following: *A firm tends to maximize its net worth.*

Equations (61)–(70) together with the fundamental assumption complete the mathematical formulation of the problem.

4.1. Optimization

The state variables are K, C_1, and A, and control variables are L, P, C_2, I_C, and I_k. Let $V(K, C_1, A, t)$ denote the maximum return function:

$$V(K, C_1, A, t) = \max W(t), \tag{71}$$

where K, C_1, A are values at t, and the maximizing is over all possible values of L, P, C_2, I_C, and I_k for $t' \geq t$. Let q_k, q_c, and q_a denote the respective adjoint variables. The Hamiltonian is given as

$$H = e^{-\gamma t}(E_g - T - I) + q_k(I_k - \delta K) + q_c(I_C - hC_1) + q_a\Phi(C_1 + C_2 - \alpha(A), P - \beta(A)). \tag{72}$$

Pontryagin's adjoint equations are

$$\dot{q}_a = -\frac{\partial H}{\partial A} = -e^{-\gamma t}(1 - T_E)G_A + q_a(\alpha'\Phi_C + \beta'\Phi_p), \tag{73}$$

$$\dot{q}_k = -\frac{\partial H}{\partial K} = -e^{-\gamma t}(1 - T_E)G_K + \delta q_k, \tag{74}$$

$$\dot{q}_c = -\frac{\partial H}{\partial C_1} = hq_c - q_a\Phi_C. \tag{75}$$

In these equations G_A and G_K denote the partial derivatives of $G(K, L, A, t)$ with respect to A and K respectively, and T_E is the partial derivative of T with respect to E_g. The prime denotes the derivative of a function of one variable, e.g., α' and β' are derivatives of α and β with respect to A. The same rule will be used to denote other derivatives and partial derivatives. The Hamiltonian is maximized by the independent variables:

$$\frac{\partial H}{\partial L} = e^{-\gamma t}(1 - T_E)(G_L - s) = 0, \tag{76}$$

$$\frac{\partial H}{\partial P} = e^{-\gamma t}(1 - T_E)(-w) + q_a\Phi_p = 0, \tag{77}$$

$$\frac{\partial H}{\partial I_C} = e^{-\gamma t}(-T_I f - f) + q_c = 0, \tag{78}$$

$$\frac{\partial H}{\partial I_k} = e^{-\gamma t}(-T_I g - g) + q_k = 0, \tag{79}$$

$$\frac{\partial H}{\partial C_2} = e^{-\gamma t}(1 - T_E)(-r) + q_a\Phi_C = 0. \tag{80}$$

From (76)

$$s = G_L. \tag{81}$$

From (77) and (80)

$$\frac{r}{\Phi_C} = \frac{w}{\Phi_p}. \tag{82}$$

The adjoint variables q_c, q_k, and q_a are obtained by solving (78)–(80):

$$q_c = e^{-\gamma t}(1 + T_I)f, \tag{83}$$

$$q_k = e^{-\gamma t}(1 + T_I)g, \tag{84}$$

$$q_a = e^{-\gamma t}(1 - T_E)(r/\Phi_C). \tag{85}$$

Differentiating the logarithms of these equations gives

$$\frac{\dot{q}_c}{q_c} = -\gamma + \frac{\dot{f}}{f} + \frac{\dot{T}_I}{1 + T_I}, \tag{86}$$

$$\frac{\dot{q}_k}{q_k} = -\gamma + \frac{\dot{g}}{g} + \frac{\dot{T}_I}{1 + T_I}, \tag{87}$$

$$\frac{\dot{q}_a}{q_a} = -\gamma + \frac{\dot{r}}{r} - \frac{\dot{T}_E}{1 - T_E} - \frac{\dot{\Phi}_C}{\Phi_C}. \tag{88}$$

Equations (73)–(75) can be rewritten as

$$\frac{\dot{q}_a}{q_a} = -\frac{G_A\Phi_C}{r} + \alpha'\Phi_C + \beta'\Phi_P, \tag{89}$$

$$\frac{\dot{q}_k}{q_k} = -\frac{1 - T_E}{1 + T_I}\frac{G_K}{g} + \delta, \tag{90}$$

$$\frac{\dot{q}_c}{q_c} = h - \frac{r(1 - T_E)}{f(1 + T_I)}. \tag{91}$$

Equating (88) with (89), (86) with (91), and (87) with (90) respectively gives

$$G_A = r\alpha' + w\beta' + \left(\gamma + \frac{\dot{T}_E}{1 - T_E}\right)\frac{r}{\Phi_C} - \frac{d}{dt}\left(\frac{r}{\Phi_C}\right), \tag{92}$$

$$(1 + T_I)\left(h + \gamma - \frac{\dot{T}_I}{1 + T_I} - \frac{\dot{f}}{f}\right)f = (1 - T_E)r, \tag{93}$$

$$(1 + T_I)\left(\delta + \gamma - \frac{\dot{T}_I}{1 + T_I} - \frac{\dot{g}}{g}\right)g = (1 - T_E)G_K. \tag{94}$$

From dynamic programming

$$q_c = V_C e^{-\gamma t}, \qquad q_k = V_k e^{-\gamma t}, \qquad q_a = V_A e^{-\gamma t}.$$

Equations (83)–(85) can be written as

$$V_C = (1 + T_I)f, \tag{95}$$

$$V_K = (1 + T_I)g, \tag{96}$$

$$V_A = (1 - T_E)(r/\Phi_C). \tag{97}$$

4.2. Interpretation of Mathematical Results

Equations (81), (82), and (92)–(97) are marginal conditions for optimum operation of the firm. Jorgenson (Ref. 8) has shown that optimal capital accumulation can mean either maximizing present value or maximizing the integral of discounted profits, and the two are the same. While Jorgenson used a simpler model, the same can be shown for the present model.

Equation (81) is the marginal productivity condition for labor services. Equation (82) gives the wage-to-rental ratio as equal to the marginal productivities ratio of programmer and computer in developing new application programs. There are two situations with large marginal productivity of computer services: (1) an increase in application programs and programming personnel and (2) introduction of a high-level language or new programming technology that increases programmer efficiency at the expense of greater computer power.

In (95)–(97) the function V is the present worth of the firm under the assumption of optimum operating conditions in the future. On the left-hand side V_C, V_K, and V_A are the marginal increases in present worth of the firm due to one additional unit increase in computer, general investment goods, and application programs. On the right-hand side are costs of such increases with tax adjusted. We note that T_I is negative and is in effect -1 if capital investment of the firm is within its investment tax credit. This is the condition under which a firm is most likely to expand its computing facilities or to invest in some other productive capital. Because application programs are not considered capital investment under tax laws, the cost r/Φ_C is adjusted for tax on earnings. The three equations state that the cost of any new investment should not exceed its added worth to the firm.

Equations (92)–(94) give criteria of a more immediate nature. The left-hand side of (94) is the cost of capital services with tax adjusted. The term $\dot{g}/g$ represents capital appreciation, while $\dot{T}_I/(1 + T_I)$ represents the depreciation of the purchasing power of investment money due to either an increase in tax or a decrease in investment tax credit. The right-hand side of (94) is the marginal productivity of capital services with tax adjusted. Equation (94) is the marginal productivity condition for capital services. Equation (93) can be interpreted in the same way with the marginal productivity of computing facilities services replaced by rental of such facilities on the right-hand side of (93).

Equation (92) equates the marginal productivity of application program services with the additional cost for running and developing such programs. The terms $r\alpha' + w\beta'$ give the cost for running additional programs, while the remaining terms are amortized cost for developing additional programs. We note that these terms are similar to the left-hand side of (94), for r/Φ_C $(=w/\Phi_P)$ is the price for developing a new application program.

Assuming that the computer manufacturers aim at increasing their sales, we make the following predictions.

The major developmental efforts of the computer manufacturers will be in new business and industrial applications and in computing facilities and software systems that make the new applications possible, economically feasible, and readily programmable.

A second developmental direction will be in diversified applications such as computer graphics, art, music, speech, health screening, diagnosis, aid to education, etc. The sum total of these developments will contribute significantly to the use of computers.

A third direction will be to develop high-level languages, devices, and software systems to improve programming efficiency and reliability.

A fourth direction will be in the development of hardware, to improve its reliability, speed, etc. While a faster and cheaper computer is not likely to improve the financial position of the computer industry as a whole, the development is unavoidable because the industry is highly competitive.

Reference

1. CHANG, S. S. L., Theory of optimum economic growth and energy policy, *Proceedings of the IEEE*, Vol. 63, pp. 414–421, 1975.
2. ARROW, K. J. and KURZ, M., *Public Investment, the Rate of Return, and Optimal Fiscal Policy*, Johns Hopkins Press, Baltimore, Md., 1973.
3. BURMEISTER, E. and DOBELL, A. R., *Mathematical Theories of Economic Growth*, Macmillan, New York, 1970.
4. ARROW, K. J., The economic implications of learning by doing, *Reviews of Economic Studies*, Vol. 29, pp. 155–173, 1962.
5. BRITTO, R., Some recent developments in the theory of economic growth: An interpretation, *Journal of Economic Literature*, Vol. 11, pp. 1343–1366, 1973.
6. CHANG, S. S. L., A theoretical study of the investment behavior of data processing users, Third World-Congress of Econometric Society, Toronto, 1974.
7. HAAVELMO, T., *A Study in the Theory of Investment*, University of Chicago Press, Chicago, 1960.
8. JORGENSON, D. W., The theory of investment behavior, *Determinants of Investment Behavior*. Edited by Robert Ferber, National Bureau of Economic Research, New York, pp. 129–155, 1967.

4

Performance Improvement of Uncertain Macroeconomic Systems

GEORGE LEITMANN AND HENRY Y. WAN JR.

1. An Alternative to Optimization

Two issues dominate the macroeconomic literature on stabilization: (1) Stabilizability: Can the economic system be steered toward a specific target and kept closely on course? (2) Optimizability: Can the stabilization program be designed according to some optimality criterion?

Stabilizability is assumed in by far the larger group of studies; certain policies are then advocated as optimal ones. Elsewhere we have reviewed this literature (as exemplified by the papers surveyed by Kendrick, Ref. 1). We believe that real-life economic systems contain certain unknown and unknowable characteristics. Consequently, no policy can claim to be optimal in either the deterministic or the stochastic sense. The design and justification of a policy must be based on an alternative approach, if one cannot hope to ascertain even the probability laws governing the happening of certain key events.

In earlier works (Refs. 2, 3), we proposed a procedure which, under certain conditions, can guarantee that (1) the controlled system is globally, uniformly, asymptotically stable relative to a target path, irrespective of the specific disturbances or uncertainties and (2) the time integral of an avoidable welfare loss is bounded from above by some finite-valued

GEORGE LEITMANN • Department of Mechanical Engineering, University of California, Berkeley, California. HENRY Y. WAN JR. • Department of Economics, Cornell University, Ithaca, New York.

function, dependent on the initial conditions of the system. In a simple example, the structure for the recommended policy was characterized, and the efficacy of monetary and fiscal measures was analyzed. In that study, we obtained uniform asymptotic stability under the recommended policy, rather than just asymptotic stability of the mean as achieved in the stochastic optimization approach.

One might inquire, as has indeed some have, what happens when the conditions we postulated fail so that the system need not be asymptotically stable. This is a matter of significant pragmatic interest, even though one can show that by introducing minor modifications the earlier procedure remains serviceable in a certain sense. This is the main thrust of the present paper.

In the rest of this section, we shall recapitulate the argument demonstrating why optimality is likely to be ill defined in reality. We shall then turn to the problem of stabilizability at a conceptual level, before we address the issue of stabilizing an unknown system that is not asymptotically stabilizable.

Decision makers in a macroeconomic problem face a more uncertain system than the usual engineering system, not only in degree but also in kind. Economic systems are affected by uncertain factors of a technical nature, for instance, the availability of unexplored mineral reserves and the severity of meteorological conditions in the coming decades, problems not confronted by the engineer in charge of a single power plant. But more fundamentally, the policymakers coexist with other agents in the same economic system. These latter individuals can have a significant impact on the system evolution, based on their free will. Three types of complications arise. We submit that no responsible policymaker can afford to ignore such factors.

First, the consequences of certain noneconomic events can severely buffet the economic system, e.g., the recurrence of a Mideast conflict and the concomitant disruption of the petroleum supply. The probability of conflict is surely not predictable by classical or Bayesian methods, yet the potential impact should certainly be anticipated in economic planning. Without reasonable expectation of probabilities, expected welfare loss is undefined.

Second, powerful interests in the economic system endeavor to anticipate the policy response of the government to current economic events. Unions may formulate their wage demands by second-guessing the vigor of antirecession fiscal policy, taking note of price hikes such an expansionary measure may cause. Large corporations may stockpile inventories if anticyclical public projects are believed to be creating shortages. This line of argument has been taken by those subscribing to the macroeconomic rational expectation hypothesis (e.g., see studies reviewed by Modigliani, Ref. 4). If the policymaker speculates about what these individuals speculate about what the policy makers speculate, ad infinitum, what can happen is at best an expectational game-theoretic equilibrium. It has nothing in common with the "passive" world that econometricians usually postulate. In fact, in a situation reminiscent of Heisenberg's uncertainty principle, the mere release of econometric forecasts may affect the confidence of the market, let alone the knowledge that powerful government policies may follow such forecasts.

Third, many advocates of stochastic optimization concede that our knowledge may be quite defective initially. Recursive procedures such as the Kalman filter are then invoked, in the hope of converging to a posterior distribution from the original prior one. In the context of macroeconomic stabilization, a trial-and-error procedure does not stop short at running and rerunning regressions. Tentative policy measures are to be implemented, with system responses sought for updating econometric estimates for the next iteration. Such a proposal, if implemented, is apt to come to grief speedily. Unlike practitioners of sequential sampling, who confront nonsentient objects in their laboratories, macroeconomic policymakers must contend with astute profit seekers, e.g., those who "raid" national currencies on the foreign exchange market. They would readily maneuver an adaptive control type of public policy to their private advantage.

Such considerations motivate us to search for an alternative approach. The heuristic feasibility of our approach seems obvious. The predator can close in on his prey, not knowing the evasive strategy of the latter, so long as he possesses timely and adequate observations, as well as sufficient maneuverability and speed. The adaptation of the pursuit–evasion differential game theory follows along quite intuitive lines: (a) Instead of aiming at the earliest guaranteed (finite) interception time, when the two objects come for the first time within capture distance, the purpose is to steer the actual trajectory to approach asymptotically the target trajectory, so that a time integral of an instantaneous loss function is guaranteed to be below some magnitude. (b) In considering a Lyapunov function for the economic system, the time derivative of such a function depends on the state, the time, the disturbances which are unknown functions of the state–time vector, and the policymaker's control which may also be a function of the state–time. If at any initial state–time and against any possible realization of the uncertain disturbance functions (whose probability distribution may be known or unknown, in advance) a stabilization policy can be found to make the time derivative of the Lyapunov function negative, then global, uniform, asymptotic stability is attained. From the economic point of view, the main interest is not so much in the system's behavior at an infinitely remote future as in the guarantee that the deviation from the target will stay below some "small" magnitude, except during an initial transition period.

Of course, one would like to know the criterion that can be used for an anticyclical policy, when asymptotic stability can no longer be assured. This relates to stabilizability.

2. On Stabilizability

Economists became acquainted with the stabilizability concepts through the works of Tinbergen (Ref. 5) and Mundell (Ref. 6). Recent studies by Aoki (Ref. 7) and others rekindled interest in this subject. The key issue is whether the economic system can be steered to reach and stay near a specific target.

The stochastic optimization approach is faced with the stabilizability question, just as our approach is. Since our approach aims at uniform asymptotic stability while the stochastic optimization approach aims only at stability in the mean, one may argue that the stochastic optimization approach is less ambitious and therefore more applicable. The issue is what stabilization policy can we propose, when asymptotic stability cannot be assured.

Such a situation may arise on two counts: (1) Deficiencies in observational capacity. The true state of the economy either cannot be accurately observed, or there is an observational delay. (2) Deficiencies in control capacity. Either the range space of the controlled system does not cover the target or the disturbances and uncertainties cannot be adequately countered by control measures.

There are certain circumstances under which policies can be devised to meet modified goals: (1) The actual path may be steered within some specific distance of the target path and kept there forever after some transitional period. (2) One may specify some unavoidable loss level such that the instantaneous excess welfare loss (above the unavoidable level) is uniformly bounded above by the smallest guaranteed level, in the long run.

Unlike engineering systems in which the simultaneous fulfillment of several requirements may determine uniquely the feasibility of a physical project, economists are interested primarily in the welfare loss due to deviation from the target. In the next section, we shall undertake a case-by-case review of various types of nonstabilizability.

3. "Noncontrollable" Systems

To focus attention on essential conceptual problems, we shall consider a sequence of very simple examples in which various earlier requirements are relaxed. Consideration of more general problems appears feasible, for instance along the lines of Leitmann (Ref. 8). As a consequence of the relaxation of certain conditions, asymptotic stability can no longer be assured but is usually replaced by ultimate boundedness in some neighborhood of the zero state. For want of a better term, we refer to such systems as noncontrollable.

3.1. The Basic Model

$x(t)$	is the level of actual effective demand,
$x^*(t)$	is the target value of effective demand,
$y(t) = x(t) - x^*(t)$	is the "inflation gap" or "deflation gap,"
$z(t)$	is the observed value of $y(t)$,
$u(t)$	is the control exerted by government,
$v(t)$	is a system disturbance or uncertainty ("noise"),
$\alpha(t)$, $\beta(t)$, $\gamma(t)$, and $\delta(t)$	are uncertain scalars.

The structural equation of the economy is

$$\dot{y}(t) = \alpha(t)y(t) + \beta(t)u(t) + v(t). \tag{1}$$

The state observation at time t allows for delay and error, namely,

$$z(t) = y(t - \gamma(t)) + \delta(t). \tag{2}$$

The uncertain quantities $\alpha(t)$, $\beta(t)$, $\gamma(t)$, $\delta(t)$, and $v(t)$ are assumed to be the values of bounded functions of y and t, continuous in y and measurable in t. In particular, we shall take the bounds to be constants:

$$\alpha(t) \in [\underline{\alpha}, \bar{\alpha}], \qquad \bar{\alpha} < 0,$$

$$\beta(t) \in [\underline{\beta}, \bar{\beta}], \qquad \underline{\beta} > 0,$$

$$\gamma(t) \in [0, \bar{\gamma}],$$

$$\delta(t) \in [\underline{\delta}, \bar{\delta}],$$

$$v(t) \in [\underline{v}, \bar{v}],$$

with $\underline{\alpha}$, $\bar{\alpha}$, etc., given constants.

We shall also suppose bounded control, say $u(t) \in [-1, 1]$.

The fact that α, β, γ, δ, and v may take on any values in the specified intervals is commonly referred to as *structural uncertainty*, *uncertainty in the policy multiplier*, *uncertainty in measurement delay*, *measurement uncertainty*, and *input uncertainty*, respectively.

In earlier work (Refs. 2, 3), we assumed the following:

(i) There is neither measurement delay nor error:

$$\gamma(t) \equiv 0 \equiv \delta(t). \tag{3}$$

(ii) There is "adequate control capacity":

$$\underline{\beta} > \max\{|\underline{v}|, |\bar{v}|\}. \tag{4}$$

We shall relax these requirements presently.

(iii) Certain "matching conditions" are met, e.g., see Gutman and Leitmann (Ref. 9). For a scalar system, these are always satisfied. More will be said about this when we consider a two-dimensional system.

(iv) The system without input and input uncertainty, i.e., with $u(t) \equiv 0 \equiv v(t)$, is asymptotically stable in that

$$\bar{\alpha} < 0. \tag{5}$$

As a matter of fact, all that is required is that $\{\alpha(t), \beta(t)\}$ is stabilizable by the addition of linear feedback. Furthermore, if (5) does not hold, the results can be extended by allowing state dependent bounds:

$$\beta(t) = \beta_1(t) + \beta_2(t)\,|y(t)|, \tag{6}$$

with

$$\beta_i(t) \in [\underline{\beta}_i, \bar{\beta}_i], \qquad i = 1, 2,$$

$$\underline{\beta}_1 > 0, \qquad \underline{\beta}_2 > \bar{\alpha}.$$

(v) The effect of control has a well-defined direction:

$$\underline{\beta} > 0. \tag{7}$$

If (7) does not hold, control may be helpful at some time and harmful at other times. The justification for using any policy must then depend on the probabilistic belief that the policy is more helpful than harmful in an uncertain world; such a belief may be unfounded so that the use of such policies must be discarded for purposes of stabilization. More will be said about this in Section 3.3.

We shall suppose that control at time t can be based only on the *observed state* $z(t)$ and, of course, on time t itself; we seek a function $p(\cdot)$: $R \times R \to R$ such that

$$u(t) = p(z(t), t), \tag{8}$$

resulting at least in ultimate boundedness of every possible system response.

Furthermore, we restrict ourselves to consideration of an instantaneous loss, say,

$$y^2(t) = \text{instantaneous loss}, \tag{9}$$

since an integral criterion might not be defined for a noncontrollable system. We note the following:

1. In the absence of control and of input uncertainty, $u(t) \equiv 0 \equiv v(t)$, the effective demand tends toward the target demand, since

$$|y(t)| \leq |y(t_0)| \exp\{\bar{\alpha}(t-t_0)\}.$$

2. In the absence of control, $u(t) \equiv 0$, the gap $y(t)$ may range indefinitely over an arbitrary interval containing

$$[y^-, y^+] \triangleq [-(\underline{v}/\bar{\alpha}), -(\bar{v}/\bar{\alpha})].$$

For that matter, if $\alpha(t) \equiv \bar{\alpha}$, $v(t) \equiv \underline{v}$, then

$$\lim_{t\to\infty} y(t) = -(\underline{v}/\bar{\alpha}),$$

and if $\alpha(t) \equiv \bar{\alpha}$, $v(t) \equiv \bar{v}$, then

$$\lim_{t\to\infty} y(t) = -(\bar{v}/\bar{\alpha}),$$

while the instantaneous loss, in the worst case, approaches

$$\lim_{t\to\infty} y^2(t) = (1/\bar{\alpha}^2) \max\{\underline{v}^2, \bar{v}^2\} \triangleq \hat{L}.$$

3. If assumptions (i)–(v) are satisfied, earlier analysis (Refs. 2, 3) indicates that one should use the following policy:

$$u(t) = \begin{cases} 1 & \text{if } y(t) < 0, \\ -1 & \text{if } y(t) > 0. \end{cases} \tag{10}$$

With this policy,

$$\begin{aligned} y(t) < 0, &\Rightarrow \underline{\alpha} y(t) + \bar{\beta} + \bar{v} \geq \dot{y}(t) \geq \bar{\alpha} y(t) + \underline{\beta} + \underline{v} > 0; \\ y(t) > 0, &\Rightarrow \underline{\alpha} y(t) - \bar{\beta} + \underline{v} \leq \dot{y}(t) \leq \bar{\alpha} y(t) - \underline{\beta} + \bar{v} < 0. \end{aligned} \tag{11}$$

Note that $\dot{y}(t)$ may take on values in some set whose members, for given $y(t)$, depend on the uncertain quantities $\alpha(t)$, $\beta(t)$, $v(t)$, and on the control $u(t)$, say, $H(\alpha(t), \beta(t), v(t), u(t); y(t))$, where $H(\cdot; y(t))$ is a set-valued function from $[\underline{\alpha}, \bar{\alpha}] \times [\underline{\beta}, \bar{\beta}] \times [\underline{v}, \bar{v}] \times [-1, 1]$ into the nonempty intervals of R.

For given initial data (y_0, t_0), the set of all solutions with

$$\dot{y}(t) \in H(\alpha(t), \beta(t), v(t), u(t); y(t)) \qquad \text{a.e.} [t_0, \infty)$$

is the *solution funnel* emanating from (y_0, t_0). The solution corresponding to

$$\dot{y}(t) = \max H(\alpha(t), \beta(t), v(t), u(t); y(t))$$

is termed the *maximal solution*, whereas that corresponding to

$$\dot{y}(t) = \min H(\alpha(t), \beta(t), v(t), u(t); y(t))$$

is called the *minimal solution.*

The situation for $\underline{v} < 0 < \bar{v}$ is sketched in Figs. 1a–c; the other cases, $0 \leq \underline{v} < \bar{v}$ and $\underline{v} < \bar{v} \leq 0$ can be analyzed in the same manner and yield similar results. Figure 1a is a generalized phase diagram, that is, one applicable to all solutions. For the uncontrolled system, $u(t) \equiv 0$, with input uncertainty $v(t)$ ranging over $[\underline{v}, \bar{v}]$ and structural parameter $\alpha(t)$ taking on its least stabilizing value, $\bar{\alpha} < 0$, gap $y(t)$ can range indefinitely over an arbitrary interval containing $[y^-, y^+]$. Thus, we assure that the eventual instantaneous loss is bounded (see Fig. 1b):

$$\lim_{t \to \infty} y^2(t) \leq \hat{L}.$$

On the other hand, use of policy (10) assures global uniform asymptotic

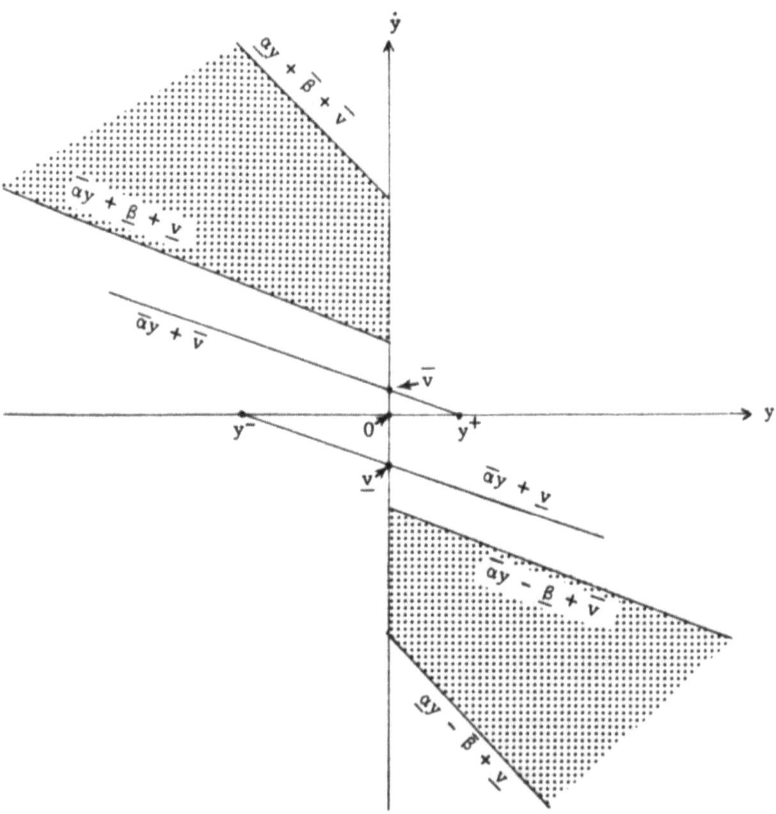

Figure 1a

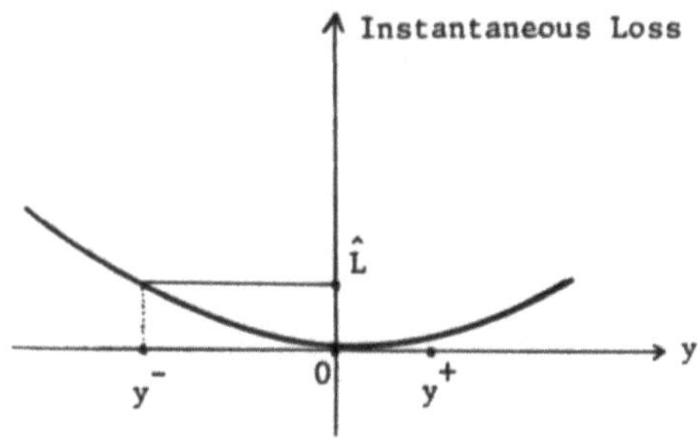

Figure 1b

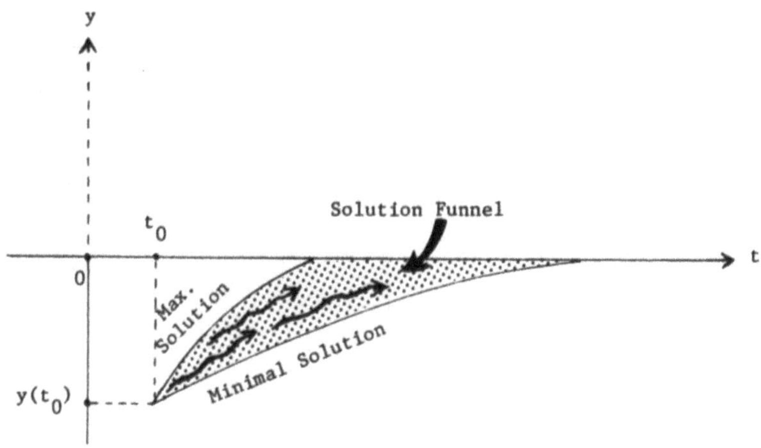

Figure 1c

stability of $y = 0$ in the presence of any possible uncertainties; however, since control is discontinuous at $y(t) = 0$, chattering is possible. Figure 1c shows a sketch of the solution funnel with control $u(t)$ given by (10).

Now we shall consider system (1), relaxing the assumptions one at a time.

Example 3.1. Inadequate Control Capacity. Suppose that inequality (4) does not hold; for instance $|\bar{v}| < \underline{\beta} < |\underline{v}|$. Consequently, if policy (10) is employed, the last inequality for $y(t) < 0$ in (11), namely,

$$\bar{\alpha}y(t) + \underline{\beta} + \underline{v} > 0$$

need not hold. The corresponding situation is sketched in Figs. 2a–c. As indicated in Fig. 2a, every trajectory with $y_0 < 0$ must eventually enter and remain within an arbitrary interval containing $[\tilde{y}, 0]$, where

$$\tilde{y} \triangleq -(\underline{v} + \underline{\beta})/\bar{\alpha} \quad \text{and} \quad y^- < \tilde{y} \leq 0.$$

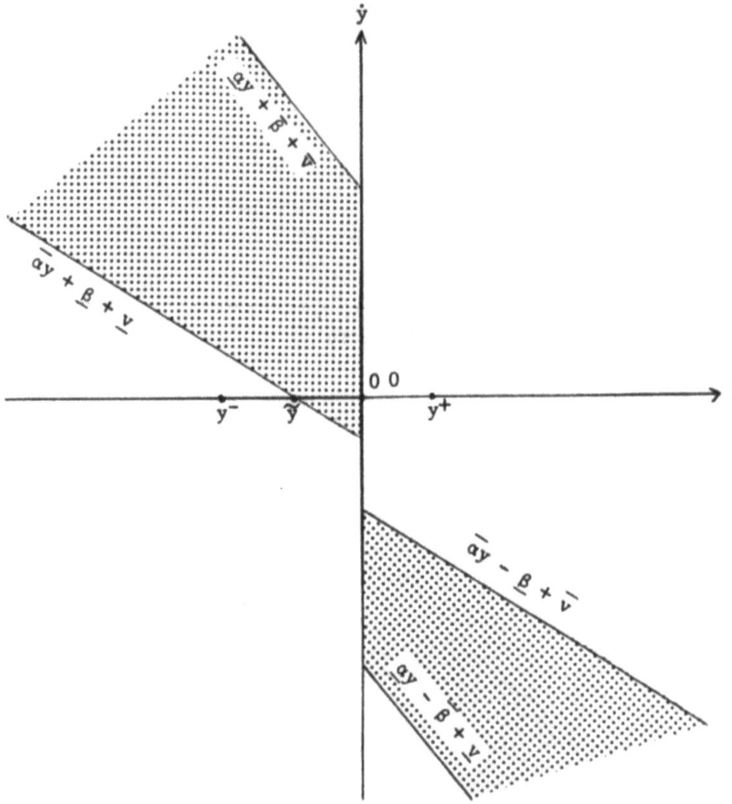

Figure 2a

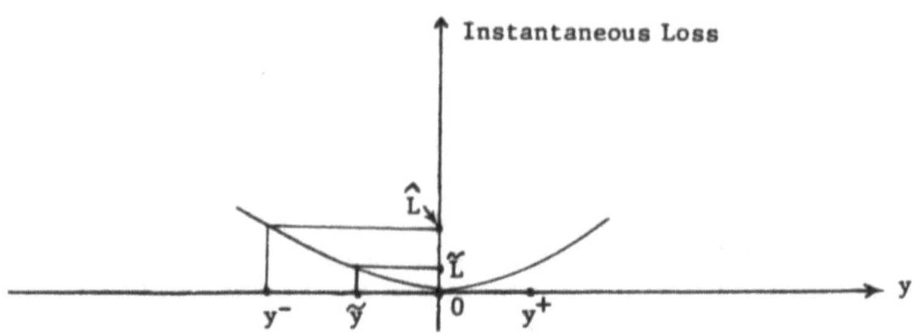

Figure 2b

As illustrated in Fig. 2b the bound of the instantaneous loss is

$$\lim_{t\to\infty} y^2(t) \leq \tilde{y}^2 = \tilde{L} < \hat{L}.$$

Figure 2c is a sketch of the corresponding solution funnel.

Thus, policy (10) does not guarantee eradication of the deflationary gap, but it does assure better performance than the uncontrolled system.

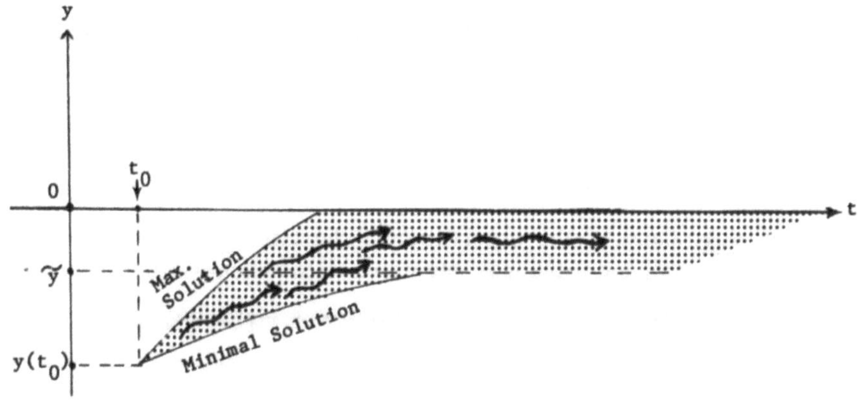

Figure 2c

Example 3.2. Error-Infested Observation. Next suppose that condition (3) does not hold. In particular, suppose that $\gamma(t) \equiv 0$ but that error, $\delta(t)$, need not vanish. Thus, when $z(t)$ is observed, $y(t)$ may range in $[z(t) + \underline{\delta}, z(t) + \bar{\delta}]$. According to policy (10), one should press on with full vigor whenever anticyclical measures are indicated. The policy employed depends entirely on the sign of $y(t)$, that is, whether the gap is inflationary or deflationary. An incorrect application of the antidote will only exacerbate the situation. Thus, if there is doubt about the symptom, one had better abstain from control.

If we take $-\underline{\delta} = \bar{\delta} \triangleq \delta$, which we may do without loss of generality by redefining the origin of $z(t)$ if necessary, then we can recommend the following policy:

$$u(t) = \begin{cases} 1 & \text{if } z(t) < -\delta, \\ 0 & \text{if } -\delta < z(t) < \delta, \\ -1 & \text{if } z(t) > \delta. \end{cases} \tag{12}$$

If $y(t) \in [-2\delta, 2\delta]$, both $z(t) < -\delta$ and $z(t) > \delta$, as well as $z(t) \in [-\delta, \delta]$, may occur. Thus, the government may either do nothing or apply the correct policy. In fact, one needs to consider four possibilities:

Case 1. $y(t) < -2\delta$.
Case 2. $-2\delta \leq y(t) \leq 0$.
Case 3. $0 \leq y(t) \leq 2\delta$.
Case 4. $y(t) > 2\delta$.

In the first and fourth case an active policy is indicated. The situation is the same as in (11); if

$$y(t) < -2\delta, \text{ then } \underline{\alpha} y(t) + \bar{\beta} + \bar{v} \geq \dot{y}(t) \geq \bar{\alpha} y(t) + \underline{\beta} + \underline{v} > 0;$$

$$y(t) > 2\delta, \quad \text{then } \underline{\alpha} y(t) - \bar{\beta} + \underline{v} \leq \dot{y}(t) \leq \bar{\alpha} y(t) - \underline{\beta} + \bar{v} < 0.$$

However, in Case 2 (Case 3), if the observation overstates the deflationary (inflationary) gap so that an active policy is employed, and if the most favorable circumstances are considered, then

$$\alpha(t) = \underline{\alpha}, \quad \beta(t) = \bar{\beta}, \quad v(t) = \bar{v} \qquad (\alpha(t) = \underline{\alpha}, \beta(t) = \bar{\beta}, v(t) = \underline{v}),$$

$$\dot{y}(t) \leq \underline{\alpha} y(t) + \bar{\beta} + \bar{v} \qquad (\dot{y}(t) \geq \underline{\alpha} y(t) - \bar{\beta} + \underline{v}).$$

On the other hand, if the observation understates the gap so that no control is exerted, and if the least favorable circumstances are considered, then

$$\alpha(t) = \bar{\alpha}, \quad v(t) = \underline{v} \qquad (\alpha(t) = \bar{\alpha}, v(t) = \bar{v}),$$

$$\dot{y}(t) \geq \bar{\alpha} y(t) + \underline{v} \qquad (\dot{y}(t) \leq \bar{\alpha} y(t) + \bar{v}).$$

An appropriate generalized phase diagram is depicted in Fig. 3.

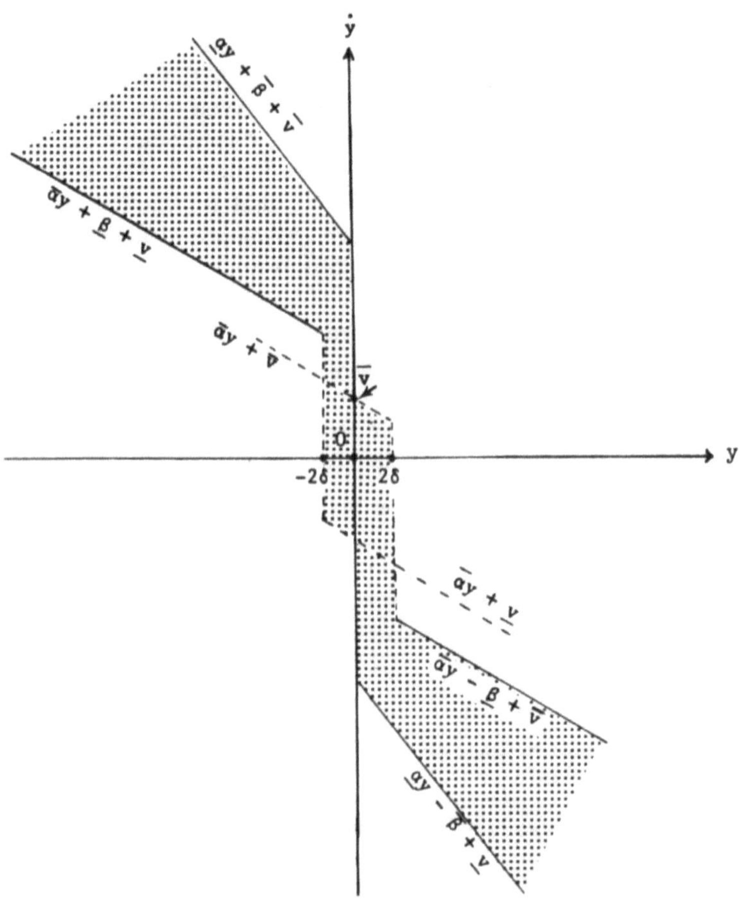

Figure 3

Thus, adoption of policy (12) guarantees ultimate boundedness of every response within an arbitrary interval containing $[-2\delta, 2\delta]$, and the bound of the instantaneous loss is

$$\lim_{t\to\infty} y^2(t) \le 4\delta^2 \triangleq \bar{L}.$$

If $2\delta < \min\{\underline{v}/\bar{\alpha}), -(\bar{v}/\bar{\alpha})\}$, then $[-2\delta, 2\delta] \subset [y^-, y^+]$ and $\bar{L} < \hat{L}$, so that use of policy (12) is superior to abstention from control.

If $2\delta > (\underline{v}/\bar{\alpha})\ (-(\bar{v}/\bar{\alpha}))$, then policy (12) guarantees a terminal interval $[y^-, 2\delta]$ ($[-2\delta, y^+]$); at worst, then, policy (12) and abstention from control assure the same terminal interval and bound of instantaneous loss. However, policy (12) is still advantageous, since it results in a more rapid convergence of the system response toward the terminal interval.

Example 3.3. Delayed Observation. Next we relax condition (3) by supposing that, while error $\delta(t) \equiv 0$, there is delay $\gamma(t) \in [0, \bar{\gamma}]$.

Loosely speaking, with delay one needs to guard against overshooting the target, that is, for instance stimulating the economy by applying control $u(t) = 1$ when there is already an inflationary gap, $y(t) > 0$, unknown to the government because of delay in information. A recommended policy is of the form

$$u(t) = \begin{cases} 1 & \text{if } z(t) < z^-, \\ 0 & \text{if } z^- < z(t) < z^+, \\ -1 & \text{if } z(t) > z^+, \end{cases} \tag{13}$$

where $[z^-, z^+]$ is the smallest interval for which overshooting is impossible. Now that $z(t) = y(t - \gamma(t))$, the question is: If one assumes that maximum delay, $\gamma(t) \equiv \bar{\gamma}$, occurs, how far ahead of reaching $z(t) = 0$ must one stop applying active control? To answer this question for any initial data we consider the maximal solution for $y(t) < 0$ [minimal solution for $y(t) > 0$] with the boundary condition $y(t^-) = 0$ ($y(t^+) = 0$) where t^- (t^+) is the time of stopping active control. Thus, for $y(t) < 0$, we consider

$$\dot{y}(t) = \underline{\alpha} y(t) + \bar{\beta} + \bar{v}, \qquad y(t^-) = 0,$$

where

$$y(t) = (\bar{\beta} + \bar{v})(e^{-\underline{\alpha}(t^- - t)} - 1)/\underline{\alpha}, \qquad t \le t^-,$$

so that

$$z^- \triangleq z(t^-) = y(t^- - \bar{\gamma}) = (\bar{\beta} + \bar{v})(e^{-\underline{\alpha}\bar{\gamma}} - 1)/\underline{\alpha} < 0. \tag{14a}$$

Similarly, for $y(t) > 0$ we find

$$z^+ \triangleq z(t^+) = y(t^+ - \bar{\gamma}) = (-\bar{\beta} + \underline{v})(e^{-\underline{\alpha}\bar{\gamma}} - 1)/\underline{\alpha} > 0. \tag{14b}$$

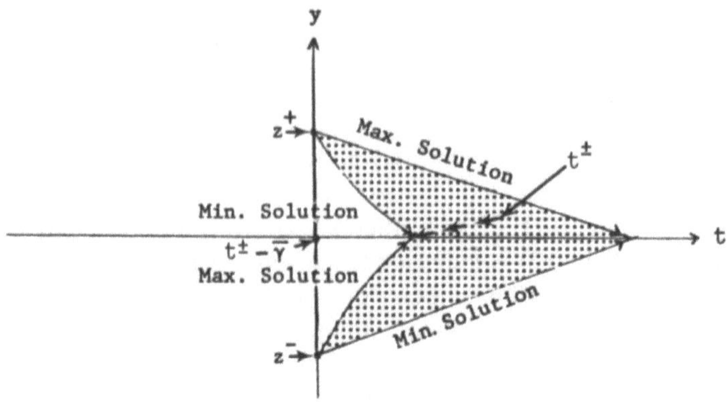

Figure 4

With policy (13), it is possible to err on the conservative side since $\gamma(t) \in [0, \bar{\gamma}]$. Thus, one can guarantee only that every trajectory will eventually enter and remain within an arbitrary neighborhood of $[z^-, z^+]$. However, this is an improvement over the null-control policy, since $[z^-, z^+] \subset [y^-, y^+]$. The instantaneous loss has a limit bounded by

$$\max\{(z^-)^2, (z^+)^2\}.$$

The corresponding solution funnel is illustrated in Fig. 4.

3.2. "Mismatched" Systems

Next we relax the matching conditions, e.g., see Gutman and Leitmann (Ref. 9). To illustrate the situation we consider a two-dimensional system described by

$$\dot{y}(t) = Ay(t) + bu(t) + v(t), \qquad y(t_0) = y_0, \tag{15}$$

where

$y(t) = \begin{pmatrix} y_1(t) \\ y_2(t) \end{pmatrix}$ is a vector of deviations from employment and price targets,

$u(t) \in [-1, 1]$ is a scalar control of monetary-cum-fiscal policy,

$v(t) = \begin{pmatrix} v_1(t) \\ v_2(t) \end{pmatrix}$ is an input disturbance with $v_i \in [-1, 1]$,

A is a constant stable 2×2 matrix,

$b = \begin{pmatrix} b_1 \\ b_2 \end{pmatrix}$ is the input vector with b_i = constant.

The instantaneous loss is now

$$y'(t)y(t) = y_1^2(t) + y_2^2(t). \tag{16}$$

To discuss the behavior of solutions of (15) we utilize a Lyapunov function approach, e.g., as in Leitmann (Ref. 8). If P is the matrix solution of the Lyapunov equation†

$$A'P + PA + I = 0, \tag{17}$$

then

$$y'Py = V(y) \tag{18}$$

is a Lyapunov function candidate since P is positive definite. The Lyapunov derivative

$$\begin{aligned} D(y, u, v) &\triangleq \nabla V(y)(Ay + bu + v) \\ &= 2y'P(Ay + bu + v) \\ &= -y'y + 2y'P(bu + v). \end{aligned} \tag{19}$$

To obtain steepest descent along a solution, we choose

$$u(t) = \begin{cases} 1 & \text{if } y'(t)Pb < 0, \\ -1 & \text{if } y'(t)Pb > 0. \end{cases} \tag{20}$$

The question now is: For what values of $\|y(t)\|$ does policy (20) guarantee negativity of the Lyapunov derivative (19)? To assure that

$$D(y, u, v) < 0 \tag{21}$$

no matter what v is, we consider

$$D(y, v, u) = -\|y\|\,[\|y\| - 2\eta P(bu + v)] \tag{22}$$

where $\eta \triangleq y/\|y\|$. Thus, we require $\|y\|$ such that

$$\|y\| - \sup_{\eta} \sup_{v} \eta' P(bu + v) > 0. \tag{23}$$

We consider two cases: control (20) and no control, $u \equiv 0$. Since, with policy (20),

$$\eta' Pbu < 0 \qquad \text{for } \eta' Pb \neq 0, \tag{24}$$

it follows that for all η such that $\eta' Pb \neq 0$,

$$\sup_{v} \eta' P(bu + v) < \sup_{v} \eta' Pv \tag{25}$$

so that

$$\underline{r} \triangleq \sup_{\eta} \sup_{v} \eta' P(bu + v) < \sup_{\eta} \sup_{v} \eta' Pv \triangleq \bar{r}. \tag{26}$$

† We may choose any positive definite constant matrix Q in place of the unitary matrix I, which we select for the sake of simplicity.

Thus, application of policy (20) is usually salutary, except in the special case for which $\sup_\eta \sup_v \eta' P(bu + v)$ occurs at η such that $\eta' Pb = 0$ so that $\underline{r} = \bar{r}$, since it guarantees a negative value of the Lyapunov derivative for all y such that

$$\|y\| > \underline{r}. \tag{27}$$

Hence, with control (20), every trajectory eventually enters and thereafter remains in an arbitrary neighborhood of the smallest Lyapunov ellipsoid

$$\{y \mid y' Py \leq \text{constant}\} \tag{28}$$

containing the ball

$$\{y \mid \|y\| \leq \underline{r}\}. \tag{29}$$

In the absence of control, this is so only for the smallest Lyapunov ellipsoid that contains the ball

$$\{y \mid \|y\| = \bar{r}\}. \tag{30}$$

The bounds of eventual instantaneous loss (16) corresponding to policy (20) and no control, respectively, are

$$\frac{\gamma_{\max}}{\gamma_{\min}} \underline{r}^2 \quad \text{and} \quad \frac{\gamma_{\max}}{\gamma_{\min}} \bar{r}^2, \tag{31}$$

where $\gamma_{\max}$ and $\gamma_{\min}$ are the maximum and minimum eigenvalues of P.

3.3. "Dubious" Instruments

Returning to the basic model of Section 3.1, we consider the following modification. In place of a single control, suppose that there are m controls

$$u^{(j)}(t) \in [-1, 1], \qquad j = 1, 2, \ldots, m,$$

so that system equation (1) is replaced by

$$\dot{y}(t) = \alpha(t) y(t) + \sum_{j=1}^{m} \beta^{(j)}(t) u^{(j)}(t) + v(t). \tag{32}$$

Suppose further that the uncertain parameters $\beta^{(j)}(t)$ are bounded; that is,

$$\underline{\beta}^{(j)}(y(t), t) \leq \beta^{(j)}(t) \leq \bar{\beta}^{(j)}(y(t), t), \tag{33}$$

where $\underline{\beta}^{(j)}(\cdot)$ and $\bar{\beta}^{(j)}(\cdot)$ are given functions, continuous in y and measurable in t.

This model is reminiscent of Turnovsky (Ref. 10), in which some of the controls sometimes may exert a dubious effect. In conformity with the adage "when in doubt, abstain," the recommended policy, in lieu of (10), is

$$u^{(j)}(t) = \begin{cases} 1 & \text{if } y(t) < 0 \text{ and } \underline{\beta}^{(j)}(y(t), t) \geq 0, \\ -1 & \text{if } y(t) < 0 \text{ and } \bar{\beta}^{(j)}(y(t), t) \leq 0, \\ -1 & \text{if } y(t) > 0 \text{ and } \underline{\beta}^{(j)}(y(t), t) \geq 0, \\ 1 & \text{if } y(t) > 0 \text{ and } \bar{\beta}^{(j)}(y(t), t) \leq 0, \\ 0 & \text{otherwise} \end{cases} \tag{34}$$

This policy is never worse, and generally better, than abstention from control.

4. Conclusion

One might question why any policy is needed if both (i) the adoption of the policy cannot guarantee asymptotic stability and (ii) even without the adoption of any policy, a stable nominal system results in ultimate boundedness of the actual one. The justification for use of the recommended policy is that it helps restrict the ultimate range of the system to a smaller neighborhood and guarantees that the instant loss will eventually have a lower bound. Even when asymptotic stability is attainable, our pragmatic interest is presumably in the system's behavior within some long but finite period.

Acknowledgment

The research of the first author was supported partially by a grant from National Science Foundation.

References

1. Kendrick, D., Applications of control theory of macro-economics, *Annals of Economic and Social Measurements*, Vol. 5, pp. 171–190, 1976.
2. Leitmann, G. and Wan, H. Y., Jr., Macro-economic stabilization policy for an uncertain dynamic economy. *New Trends in Dynamic Systems Theory and Economics*, Springer-Verlag, Vienna–New York, 1978.
3. Leitmann, G. and Wan, H. Y., Jr., A stabilization policy for an economy with unknown characteristics, *Journal of Franklin Institute*, Vol. 306, No. 1, pp. 23–33, 1978.
4. Modigliani, F., The monetarist controversy or should we forsake stabilization policies? *American Economic Review*, Vol. 67, pp. 1–19, 1977.
5. Tinbergen, J., *On the Theory of Economic Policy*, North Holland, Amsterdam, 1952.
6. Mundell, R. A., The nature of policy choice, *Banco Nazionale del Lavoro Quarterly Review*, Vol. 26, 16, pp. 262–284, 1963.

7. Aoki, M., On a generalization of Tinbergen's condition in the theory of economic policy to dynamic models, *Review of Economic Studies*, Vol. 42, pp. 293–296, 1975.
8. Leitmann, G., Guaranteed ultimate boundedness for a class of uncertain linear dynamical systems, *Proceedings of the Third Kingston Conference on Differential Games and Control Theory*, June 1978, Marcel Dekker, New York, 1979.
9. Gutman, S. and Leitmann, G, Stabilizing feedback control for dynamic systems with bounded uncertainty, *Proceedings of IEEE Conference on Decision and Control*, 1976.
10. Turnovsky, S. J., Optimal stabilization for deterministic and stochastic linear systems, *Review of Economic Studies*, Vol. 40, pp. 79–95, 1973.

5

Stochastic Games, Oligopoly Theory, and Competitive Resource Allocation

MARTIN SHUBIK AND MATTHEW J. SOBEL

1. Introduction

An oligopolistic market is one with only a few firms that supply the commodity being purchased. Oligopoly theory, until recently, evolved without regard to the institutional details encountered in specific markets and without addressing the role played by time. Oligopoly models were treated statically, or at best, conversationally dynamically. However, dynamic oligopoly models have been analyzed with increasing frequency in recent years and some of these analyses are responsive to institutional details.

Here we compare the literature on dynamic models of oligopoly with our intepretation of the objectives of oligopoly theory. We use discrete time-sequential games, sometimes called "stochastic games," as a canonical form in which to discuss the issues. The stochastic game model encompasses many interesting oligopoly models and it seems to offer an appropriate level of generality to address research needs. Incidentally, we do not believe that there is any importance to economic theory associated with the distinction between continuous and discrete time models, i.e., between stochastic games and differential games. In principle, the discus-

---

MARTIN SHUBIK • Cowles Foundation, Yale University, New Haven, Connecticut.
MATTHEW J. SOBEL • College of Industrial Management, Georgia Institute of Technology, Atlanta, Georgia.

sion could be couched in terms of continuous time models instead of stochastic games.

We believe that the primary objective of oligopoly theory is to provide an understanding of pricing and resource allocation over time in large firms, and the consequent market behavior of such firms. We begin by enumerating some issues involving dynamics that are inherent in this goal. We sketch a general stochastic game model and use the model to address the task of constructing satisfactory dynamic oligopoly models. This effort in part becomes a specification of research needs and opportunities in stochastic game theory and oligopoly theory. At this stage in the development of both subjects, it is useful to identify problems rather than only describe past accomplishments. Section 5 cites specific recent results, and Section 6 presents a blending of problems in optimization and survival.

2. Dynamical Issues in Oligopoly

What are some characteristics of the dynamics of pricing and resource allocation in large firms, and their consequent market behavior? Here, we mention three kinds of problems. First, there are the dynamics of the composition of the set of firms in the market. This "entry and exit" problem in oligopoly is the subject of current research but most analyses have either been static or ignored the multiplicity of "players" in such games. A closely related problem is how to distinguish "competition among the few" from "competition among the many." The modeling issue is how large must a market become in order for gamelike individual behavior to become unimportant. Second, in a given oligopolistic market, why do prices fluctuate as they do? In most markets, the prices fluctuate more slowly than the prices of the factors comprising the inputs in the production process. This phenomenon of "sticky prices" is widely recognized but has hardly been analyzed in a dynamic oligopoly model. Lastly, in some oligopolistic markets, there is one firm that acts as a leader in changing the price level. Why? Why is there price leadership behavior in some markets but not in others? Why might a firm act passively as a follower under some conditions but bolt the pack under other conditions?

Another collection of dynamical issues concerns the role of information in market behavior. How do firms tacitly communicate their objectives, strategies, and threats to one another? How do divisions of a large firm communicate with one another so that their decentralized actions are mutually supportive of the overall goals of the firm. This is the general problem of managerial control. Furthermore, how do accounting conventions affect firm and market behavior? Technically, this question can be posed in terms of alternative aggregations of information. Lastly, what are the effects of imperfections in information, particularly those due to delays in transmission of information? Little progress has been made on a general treatment of this last issue and the prior one has been analyzed in some detail only in static models (Ref. 1).

What are the effects of market size? Most facets of this issue are not particularly dynamical in nature but we should know how to analyze them in dynamic models. As one example, what is the effect on product quality of the number and size of firms that are competing? An issue that is primarily dynamical is the dependence of the number of firms in the market upon the time rate at which information spreads, and vice versa.

Preference structures have been treated somewhat incidentally in oligopoly theory. Important research on intertemporal preference orderings is currently being done (Ref. 2) for models of individual decision-making over time. Comparable investigations of dynamic multiperson decision models have not yet begun. The situation becomes even more complicated if we construct "behavioral models of the firm" (Refs. 3 and 4) that discard the notion of a single monolithic "decision maker" making all the decisions in each firm. The models in "team theory," for example, can be construed as noncooperative games among players having the same preference ordering over outcomes but differing in the information and the actions available to each. We have yet to see an investigation of sequential models of this kind. Lastly, the dynamic oligopoly models analyzed thus far are predicated on a scalar objective such as each firm's discounted operating profit. However, various economists argue that managers in firms behave as if they were maximizing vector objectives. Components other than profit might include rate of growth in sales, number of employees, share of the market, and survival. An important first step has been taken in the analysis of sequential games with vector payoffs (Ref. 5) but this general theory has yet to be applied to a dynamic oligopoly model.

We now turn to some issues of constructing satisfactory dynamic oligopoly models. The canonical form of a general stochastic game will be useful for that purpose. The next section briefly defines a stochastic game and enumerates some notions on the "solution" of such a model.

3. Stochastic Games

Let I be a set of *players*, S a set of *states*, and A_s^i a set of *actions* available to player $i \in I$ when the process is in state $s \in S$. These sets are assumed to be nonempty. The composite action of all the players, when the process is in state s, must be an element of $C_s = \times_{i\in I} A_s^i$. We write $a = (a^i) \in C_s$. An outcome of a stochastic game is a sequence $s_1, a_1, s_2, a_2, \ldots$, where $a_t \in C_{s_t}$ for all t. Let $W = \{(s, a): a \in C_s, s \in S\}$.

The dynamics are determined by the decision rules used by the players to choose their actions and by a collection $\{q(\cdot \mid s, a): (s, a) \in W\}$ of probability measures on B_s, the Borel subsets of S. For any period t and $H \in B_s$, if $s_t = s$ and $a_t = a$ then $q(H \mid s, a)$ is the probability that $s_{t+1} \in H$.

A two-person zero-sum matrix game is a special case of a stochastic game where S is a singleton. It is easy to see that in such a game, in general, one may wish to admit randomized strategies. This complicates the measurability and integrability issues that are already embedded in the one-person stochastic game, namely, the Markov decision process. Our exposition suppresses these issues, which are the subject of some current research on stochastic games. The interested reader should read the fine survey by Parthasarathy and Stern (Ref. 6) and the excellent recent paper by Whitt (Ref. 7).

With the preceding caveat, let π^i denote the set of player i's nonanticipative decision rules (including rules that are history dependent and randomized) for choosing a_t^i, for each t, on the basis of the outcome to date, namely, $s_1, a_1, \ldots, s_{t-1}, a_{t-1}, s_t$. The *stationary* policies are particularly interesting decision rules. Let D_s^i be the set of probability measures on the Borel subsets of A_s^i. An element of D_s^i can be used to choose a randomized action when the game is in state s. Let $\Delta^i = \times_{s\in S} D_s^i$, $\Delta = \times_{i\in I} \Delta^i$, and $\pi = \times_{i\in I} \pi^i$. An element of π is a policy. A policy $\gamma \in \pi$ is stationary if there exists $\delta \in \Delta$ such that $\gamma = (\delta, \delta, \delta, \ldots)$ so $a_t = \delta(s_t)$ for all t. We write $\gamma = \delta^\infty$ in this case. Let Δ^∞ denote the subset of stationary policies in π. Finally, it is convenient to represent any $\gamma \in \pi$ as (γ^i, γ^{-i}), where $\gamma^{-i} \in \prod_{j\neq i} \pi^j$.

There has been some research on the ergodic properties of $\{(s_t, a_t)\}$ induced by stationary policies (cf. Refs. 8 and 9) but most literature concerns real-valued payoff functions. Let $r^i(s, a)$ denote the (expected) immediate reward to player i in any period t if the state s_t is s and the composite action a_t is $a \in C_{s_t}$. Let β_i be player i's single-period discount factor and let

$$\begin{aligned} V^i(\gamma \mid s) &= \sum_{t=1}^{\infty} \beta_i^{t-1} r^i(s_t, a_t) \\ v^i(\gamma \mid s) &= EV^i(\gamma \mid s) \end{aligned} \tag{1}$$

denote the total discounted payoff and its expectation when $\gamma \in \pi$ is the policy and $s_1 = s$ is the initial state. Some literature concerns the average payoff per period rather than the discounted payoff but the latter is more appropriate for oligopoly and other open (or partial equilibrium) economic models.

A policy $\gamma \in \pi$ is said to be an *equilibrium point relative to* $H \subseteq S$ iff

$$v^i(\gamma \mid s) = \sup \{v^i(\rho, \gamma^{-1}): \rho \in \pi^i\}, \qquad s \in H, \quad i \in I. \tag{2}$$

We say simply that γ is an *equilibrium point* if it is an equilibrium point relative to S. An equilibrium point relative to H is noncollusively optimal for every initial state in H and every player. Shapley (Ref. 10), in a magnificent early paper, established existence of an equilibrium point amongst stationary policies in a two-player model with $\bigcup_{s \in S} A_s^i$ being a finite set for each player and $r^i(\cdot, \cdot) + r^2(\cdot, \cdot) \equiv 0$. More general existence results concern nonzero sum games played by more than two players (Refs. 6, 7, 11, 12). Also, Henig (Ref. 5) has recently established the existence of an equilibrium point for games where $r^i(\cdot, \cdot)$ is vector-valued.

4. Modeling Issues

We have observed that the extent theory fails to explain why price leadership occurs and which firms are likely to be followers while others are leaders. More generally, there is no satisfactory sequential analogue to "cooperative" theory for static games. The primitive element in most of the cooperative static theory is the coalition, namely, a subset of the players who join together for mutual benefit. However, a satisfactory dynamic theory must admit changes in coalition composition as time passes, but the present theory does not include this feature.

Another modeling issue stems from the embarrassment of riches provided by the size of the set of equilibrium points. It is known that the size of this set increases as information conditions in a game proliferate (cf. Ref. 13). Therefore, oligopoly models that strive to include the design of information systems and managerial control may induce distressingly many equilibrium points. The problem is to decide which one, more generally which subset, is the appropriate object for analysis. We believe that the solution to this problem should vary depending upon the context which motivates the model. In other words, behavioral considerations and institutional details should direct our definition of "the appropriate object for analysis."

It has already been mentioned that no satisfactory canonical model exists to analyze the effects of imperfections in information due to delays in transmission. Indeed, the intricacy of the analysis in a relatively simple case analyzed by Scarf and Shapley (Ref. 14) is alarming. We doubt that a Bayesian approach is appropriate here although one of us has explored this elsewhere (Ref. 8). The extant theory of stochastic games would oblige us to assume that each firm knows the prior distribution held by every other firm.

Careful modeling of many industries leads to the explicit inclusion of bankruptcy conditions in a model. Such conditions exemplify "exit fees" in the class of Markov decision processes called *stopping problems.* There are several interesting qualitative results concerning the structure of optimal policies in stopping problems. As yet, there is no comparable theory for "stopped sequential games." The development of such a theory might have a payoff for oligopoly theory by offering a deeper understanding of the effects of alternative bankruptcy laws and the dynamics associated with the entry and exit of firms from an industry. One of us has suggested a class of "games of economic survival" to pick up the ruin posibilities (cf. Section 6; Refs 15 and 16).

This list of modeling issues is necessarily brief and we have not discussed some pertinent material. Aumann (Ref. 17) has developed results for "supergames." A supergame is a sequence of static games in which the nature of the static games is not contingent on players' past actions. The case of the same static game at each point (in the sequence) has been investigated more than any other. This case is a stochastic game with $|S| = 1$. Friedman (Ref. 18) analyzes this case in oligopoly models. He focuses on "reaction function" strategies; each player's present decision is contingent on the opponent's preceding decision. Such decision rules induce a stochastic game in which S is the set of player's possible single-game decisions. Recently, Rosenthal (Ref. 19) has investigated sequences of games with varying opponents. His point of view may be useful for construction of dynamic models of entry and exit in oligopoly. Shefrin (Ref. 20) has interesting results for dynamic market games with incomplete information.

5. Specific Results

Stochastic game models of oligopoly, even with the limitations enumerated above, are forbiddingly complex. Nevertheless, some progress has been made either by reducing the potential complexity or by building a model for a particular kind of industry and then posing correspondingly special questions. First, we discuss the reduction of complexity.

Stochastic game models are difficult to analyze because the number of players is greater than one, so the players interact with one another, and because the game process extends over time and each player indulges in a variety of intertemporal tradeoffs. Several writers (see Ref. 21 and references therein) have suppressed the complexity owing to the interaction of firms by analyzing models of leader–follower behavior where the identities of the leader and followers are known at the outset. The problem is then the selection of an optimal dynamic policy by the leader and this latter problem is a (one-person) Markov decision process which is much less complex than a stochastic game.

Another suppression of complexity has been obtained by preserving a multiplicity of players (firms) but reducing the original dynamic game to a static game. Specifically, an equilibrium point of a stochastic game is said to be *myopic* if it consists of the *ad infinitum* repetition of an equilibrium point of a static game. The principal sufficient conditions (Ref. 22), satisfied by various dynamic oligopoly models, are:

> for each $i \in I$ and $(s, a) \in W$, $r^i(s, a)$ depends additively on the state s and action a, i.e., there are functions K^i and L^i such that
>
> $$r^i(s, a) = K^i(a) + L^i(s); \tag{3a}$$
>
> transition probabilities depend on the actions taken but not on the state from which transition occurs, i.e.,
>
> $$q(H \mid s, a) = p(H \mid a) \qquad \text{for all} \quad H \in B_S \quad \text{and} \quad (s, a) \in W; \tag{3b}$$
>
> suppose the static game Γ, defined below, has an equilibrium point a_* in pure strategies and let $S^* = \{s: s \in S, a^* \in C_s\}$. Then $p(H \mid a^*) = 1$ for all $H \in B_s$ having $S^* \subseteq H$, i.e., if $s_1 \in S^*$ then a^* is repeatable *ad infinitum* (with probability one). (3c)

Let $\xi(a)$ be a random variable with the measure $p(\cdot \mid a)$ in (3b). Then I is the set of players in the static Nash game Γ, player i's payoff function is $K^i(a) + \beta_i E[L^i(\xi(a))]$, $a \in \bigcup_{i \in I} A^i$, and player i has available the set of moves $A^i = \bigcup_{s \in S} A^i_s$. If a_* is randomized then there is an assumption comparable to (3c). It follows from (3) that $a_t = a_*$ for all t is an equilibrium point [in the stochastic game sense of (2)] relative to S^*.

Numerous Markov decision processes in the literature satisfy (3) but the myopia of their optimal policies was either overlooked or deduced by special and sometimes intricate arguments. Also, various oligopoly models satisfy (3). Kirman and Sobel (Ref. 23) assume that firms make production and pricing decisions each period and that they have linear production costs and arbitrary single-period inventory-related costs. The oligopoly model in Sobel (Ref. 24) focuses on advertising decisions. It is assumed there that each firm's demand each period is a random variable whose distribution depends on all firms' "goodwill." The goodwill is an exponentially weighted moving average of past amounts spent on advertising. Myopia has been applied to other oligopoly models where the competition involves expenditures on research and development, expansion of capacity, and the harvesting of interacting fish species in a coastal fishing industry (Ref. 25).

6. Games of Economic or Social Survival

The goals of profit maximization or cost minimization are present in many economic models. Games of survival, as characterized by Shubik (Ref. 15), stress the binary outcomes of survival or nonsurvival. Yet in many social, ecological, and economic processes the goals include both survival and optimization of the quality of life for the survivor.

Below we describe a general class of games of social or economic survival. With notational changes, they can be recast as stochastic games.

An n-person game of economic survival is described as follows by

$$\alpha^i(t);\ W_1^i;\ B_1^i;\ S_1^i;\ L_1^i;\ \Psi_1^i;\quad \text{and}\quad V^i,$$

where

$\alpha^i(t)$, $i = 1, \ldots, n$, are the single-period *payoffs* faced by the players at time t. The payoffs depend on actions described below and will in general depend upon time.

W_t^i, $i = 1, \ldots, n$ and $t = 1, 2, \ldots$, are the *wealths* of the players at the start of time t. Initial assets W_1^1 and W_1^2 are given as parameters.

B^i are the *ruin conditions*, bankruptcy levels or "absorbing barriers," for the player; i.e., if the assets (or strength or viability) of a player i drop to below B^i, that individual is out of the game.

S^i are the *survival values*; i.e., if individual i is the sole survivor in the game, S^i is the present value of the remaining one-person game.

L^i are the *liquidation values*. If an individual i is ruined, he may still have residual assets at the point of ruin. The value of these assets is given by L^i.

Ψ^i are *discount factors*. We assume that each individual has a discount factor on future consumption. The Ψ^i could be dependent upon the age of the individual, thus reflecting life-cycle considerations.

T is the *time* at which bankruptcy (if any) occurs:

$$T = \inf\{t: (W_t^1, W_t^2) \not> (B^1, B^2)\}.$$

V^i is the *payoff function* to player i. It cannot be fully specified until the strategies and the relationship among income, consumption, and survival are specified.

Two versions are given, the game where pure survival is the goal and the game where the maximization of expected discounted consumption [or utility $\phi^i(\cdot)$ of consumption] is optimized.

At the start of any time t an individual i has W_t^i. A Markov strategy by an individual i is a plan for the selection of an investment amount x_t^i and a consumption amount b_t^i dependent upon $W_t = (W_t^1, \ldots, W_t^n)$.

$$W_{t+1}^i = \xi_t[W_t^i + \alpha^i(x_t^1, \ldots, x_t^n) - b_t^i - x_t^i],$$

where $0 < x_t^i + b_t^i < W_t^i$ and $\xi_1, \xi_2, \ldots$ are independent and identically distributed random variables.

For the game of *pure survival*,

$$V^i = \begin{cases} 1 & \text{if } W_t^i > B^i \quad \text{for} \quad t = 1, 2, \ldots, \\ 0 & \text{otherwise.} \end{cases}$$

For the game of *pure consumption optimization*.

$$V^i = \begin{cases} \sum_{t=1}^{\infty} (\Psi^i)^{t-1}\phi^i(b_t^i), & W_t^i > B^i, t = 1, \ldots, \infty, \\ & \text{and game continues indefinitely} \\ \sum_{t=0}^{T} (\Psi^i)^{t-1}\phi^i(b_t^i) + (\Psi^i)^T S^i, & W_t^i > B^i, t = 1, \ldots, T, \\ & \text{and game continues until } T \\ \sum_{t=1}^{T} (\Psi^i)^{t-1}\phi^i(b_t^i) + (\Psi^i)^T L^i, & W_t^i > B^i, t = 1, \ldots, T-1, \\ & W_T^i \le B^k, \text{ and game continues} \\ & \text{until } T-1. \end{cases}$$

For either criterion, let J indicate which player (if either) is ruined:

$$J = \begin{cases} 0 & \text{if} \quad T = \infty, \\ j & \text{if} \quad W_T^j \le B^j. \end{cases}$$

Then, for either criterion, consider the vector (V^1, V^2, T, J). The adoption of a policy by the players induces a (joint) probability distribution of this vector and we may compare the distributions induced by alternative policies. An obvious comparison is according to $v^i = E(V^i)$. For example, in the pure survival criterion, $v^i = P\{V^i = 1\} = P\{J \ne i\}$. Although we shall not pursue the matter here, the two criteria can be treated in a unified manner by first defining an appropriate stochastic game.

6.1. A "Guns or Butter" Example

A two-player example illustrates the tradeoff between consumption and survival. Let b_t and x_t denote player 1's consumption and investment in period t and let d_t and y_t denote the same amounts for player 2.

Let

$$\alpha^1(t) = \begin{cases} A\dfrac{x_t}{x_t + y_t} & \text{if} \quad x_t + y_t > 0, \\ 0 & \text{if} \quad x_t = y_t = 0, \end{cases}$$

$$\alpha^2(t) = \begin{cases} A\dfrac{y_t}{x_t + y_t} & \text{if} \quad x_t + y_t > 0, \\ 0 & \text{if} \quad x_t = y_t = 0. \end{cases}$$

Let $W_1^1 = M$; $W_1^2 = m$; $B^1 = B^2 = L^1 = L^2 = 0$, $S^1 = S^2 = A/(1-\Psi)$; $\Psi^i(c) = c$; $\Psi^1 = \Psi^2 = \Psi$, where $0 < \Psi < 1$. Say $b_t^1 = b_t$ and $b_t^2 = d_t$.

If the goals are pure survival then $x_t = y_t = b_t = d_t = 0$ gives $M_{t+1} = \xi_t M_t > 0$, $m_{t+1} = \xi_t m_t > 0$, where $\xi_t > 0$ is a random variable.

If the goals are maximization of expected consumption then if a solution with joint survival is feasible, player 1 wishes to

$$\max_{b_t, x_t} E \sum_{t=1}^{\infty} (\Psi^i)^{t-1} b_t, \qquad b_t \geq 0, \qquad x_t \geq 0$$

subject to $0 \leq b_t$, $0 \leq x_t$,

$$b_t + x_t \leq M_t$$

and

$$M_{t+1} = \xi_t \left[M_t - b_t + A\left(\frac{x_t}{x_t + y_t}\right) - x_t \right]$$

and player 2 wishes to

$$\max_{d_t, y_t} E \sum_{t=1}^{\infty} (\Psi^i)^{t-1} d_t$$

subject to $0 \leq d_t$, $0 \leq y_t$,

$$d_t + y_t < m_t$$

and

$$m_{t+1} = \xi_t \left[m_t - d_t + A\left(\frac{y_t}{x_t + y_t}\right) - y_t \right].$$

Let $\mu = E[\xi_t]$. If $\Psi\mu \leq 1$ and $P\{\xi \geq \Psi\mu/2\} = 1$ (for which $P\{\xi \geq \frac{1}{2}\} = 1$ is sufficient) then we can show that there is a myopic equilibrium point with respect to $\{(M_1, m_1): (M_1, m_1) \geq \Psi\mu A/4(1, 1)\}$ given by

$$\begin{aligned} x_t &= y_t = \mu A \Psi/4, && \text{for } t = 1, 2, \ldots, \\ b_t &= d_t = A(2\xi_{t-1} - \Psi\mu)/4, && \text{for } t = 2, 3, \ldots, \\ b_1 &= M_1 - \mu A \Psi/4, && d_1 = m_1 - \mu a \Psi/4 \end{aligned}$$

and expected payoffs are

$$\begin{aligned} v^1 &= M_1 - A\mu\Psi/4 + \sum_{t=2}^{\infty} \Psi^{t-1}(2\xi_{t-1} - \Psi\mu)A/4 \\ &= M_1 + A\mu\Psi/[4(1 - \Psi)], \\ v^2 &= m_1 + A\mu\Psi/[4(1 - \Psi)]. \end{aligned}$$

If $P\{\xi_1 < \Psi\mu/2\} > 0$, or $(M_1, m_1) \not\geq \Psi\mu A/4(1, 1)$, or $\Psi\mu > 1$ then the analysis and solution become complicated.

The "net earnings" or gains from competition are given by the terms

$$A\frac{x_t}{x_t + y_t} - x_t.$$

These portray the resource struggle or the "battle conditions." The payoffs involve only the b_t, i.e., the resources drawn off for consumption.

ACKNOWLEDGMENT

The preparation of this work was partially supported by NSF grants ENG76-46156 and SOC78-T5730. This work relates to Department of the Navy Contract N00014-77-C-0518 issued by the Office of Naval Research under Contract Authority NR-047-006. However, the content does not necessarily reflect the position or the policy of the Department of the Navy or the Government, and no official endorsement should be inferred.

References

1. MARSCHAK, J. and RADNER, R., *Economic Theory of Teams*, Yale University Press, New Haven, 1972.
2. KREPS, D. M. and PORTEUS, E. L., Temporal resolution of uncertainty and dynamic choice theory, *Econometrica*, Vol. 46, pp. 185–200, 1978.
3. CYERT, R. M. and MARCH, J. G., *A Behavioral Theory of the Firm*, Prentice Hall, Englewood Cliffs, N.J., 1963.
4. WILLIAMSON, O. E., *Markets and Hierarchies*, Free Press, New York, 1975.
5. Henig, M., *Multicriteria Dynamic Programming*, Yale University, Ph.D. Thesis, 1978.
6. PARTHASARATHY, T. and STERN, M., On Markov games, *Differential Games and Control Theory*, Vol. 2, Edited by E. Roxin, P. T. Liu, and R. Sternberg, Marcel Dekker, New York, pp. 1–46, 1977.
7. WHITT, W. Representation and approximation of noncooperative stochastic games," *J. Siam*, to appear.
8. SANGHVI, A. P. and SOBEL, M. J., Bayesian games as stochastic processes, *International Journal of Game Theory*, Vol. 5, pp. 1–22, 1976.
9. SANGHVI, A. P., Sequential games as stochastic processes, *Stochastic Processes and Their Applications*, Vol. 6, pp. 323–336, 1978.
10. SHAPLEY, L. S., Stochastic games, *Proceedings of the National Academy of Sciences, USA*, Vol. 39, pp. 1095–1100, 1953.
11. ROGERS, P. D., *Nonzero-Sum Stochastic Games*, Operations Research Center, University of California, Berkeley, Report ORC-69-8, 1969.
12. SOBEL, M. J., Noncooperative stochastic games, *Annals of Mathematical Statistics*, Vol. 42, pp. 1930–1935, 1971.
13. DUBEY, P. and SHUBIK, M., Information conditions, communications, and general equilibrium, *Mathematics of Operations Research*, 1979, in press.
14. SCARF, H. E. and SHAPLEY, L. S., Games with partial information, *Contributions to the Theory of Games*, Volume III, Edited by M. Dresher, A. W. Tucker, and P. Wolfe, Princeton University Press, Princeton, N.J., pp. 213–230, 1957.
15. SHUBIK, M., *Strategy and Market Structure*, Wiley, New York, 1958.

16. Shubik, M. and Thompson, G. L., Games of economic survival, *Naval Research Logistics Quarterly*, Vol. 6, pp. 111–123, 1959.
17. Aumann, R. J., Acceptable points in general cooperative n-person games, *Contributions to the Theory of Games. IV*, Edited by R. D. Luce and A. W. Tucker, Annals of Mathematics Study 40, Princeton University Press, Princeton, New Jersey, pp. 287–324, 1959.
18. Friedman, J. W., *Oligopoly and the Theory of Games*, North-Holland, Amsterdam, 1977.
19. Rosenthal, R. W., Sequences of games with varying opponents, *Econometrica*, 1979, in press.
20. Shefrin, H. M., Bayesian convergence in market games with incomplete information, *Proceedings of the Third Kingston Conference on Differential Games and Control Theory, 1978*, Marcel Dekker, to appear.
21 Lippman, S. A., Optimal pricing to retard entry, *Review of Economic Studies*, Vol. 44, p. 32, 1977.
22. Sobel, M. J., Myopic equilibrium points in stochastic games, unpublished manuscript, 1978.
23. Kirman, A. P. and Sobel, M. J., Dynamic oligopoly with inventories, *Econometrica*, Vol. 42, 279–287, 1974.
24. Sobel, M. J., "Sequential Marketing Games," unpublished manuscript, 1977.
25. Sobel, M. J., "A Stochastic Game Model of a Regional Fishery," unpublished manuscript, 1979.

Part II

Dynamical Theory of the Economics of Extractive Natural Resources

Naturally growing extractive resources are renewable or nonrenewable, depending on whether they are regenerated at significant rates over time. Examples of renewable resources are fisheries, forests, and ground water; of nonrenewable resources, hard minerals and the hydrocarbon fuels.

While Hotelling applied the calculus of variation to the nonrenewable resource problem as early as 1931, systematic though simplistic treatment of the inherently dynamic model problems did not begin until the late 1950s. Recent years have witnessed a flourishing body of economic literature applying modern control theory to extractive resources.

The recent energy crisis and the extension of jurisdiction over marine fish stocks have given rise to numerous concerns in the public sector. Principal among the concerns are the resource scarcity and exploration, production efficiency and distribution under alternative market structures and institutional arrangements, and optimal policies to achieve various social goals. The five papers in this part treat various aspects of these important issues.

The essence of the extractive resource problem is a differential equation of the form

$$\dot{x} = F(x) - h(t), \tag{1}$$

where $x(t)$ is the size of resource stock at time t, $F(\cdot)$ is the rate of natural growth, and $h(t)$ is the rate of extraction. For nonrenewable resources, the growth rate is negligible and $F(\cdot) = 0$. The typical problem is to optimize some functional subject to (1) and examine the intertemporal path of $h(t)$ and possibly other control variables.

As Clark has pointed out, harvesting of renewable resources is a problem of managing a flow of goods and can be studied on a capital-theoretic framework. In this respect, the problem is analogous to that of economic growth. Stability of a steady-state solution is often one of the most crucial points in the analysis of such a problem, because it pertains to the question of whether a species could be sustained at a certain level under some harvesting policy. Clark's book on optimal management of renewable resources contains a comprehensive treatment of many aspects of the problem.

In this part, Clark extends his results on optimal harvesting policies for sole ownership of a fishery resource property to the case of multi-ownership of a common property. The problem is formulated as an N-person nonzero-sum differential game. Under the assumption that competition exists among all agents, a Nash equilibrium solution is obtained. With such a solution, the more "efficient" agent may be able to eliminate his competitors if the level of stock is reduced to an appropriate level. He may even be able to operate at the optimum level of sole ownership if the effective advantage is large. A more general model of restricting each agent's input level is also discussed. With relatively simple models and mathematical analysis, this paper presents and discusses many economic issues in the exploitation of common property fishery resources by using the game-theoretic approach.

For an exhaustible resource, the differential equation describing the process of exploitation is simpler, since $F(\cdot) = 0$. But despite the simplicity of the equation, the problem is often complicated by many factors. For one thing, the amount of reserve is usually not exactly known and remains to be determined as exploitation goes on. This requires stochastic and adaptive control techniques. When stochastic models are used, the firm's policies are very much dependent on its attitude toward risk (risk taking or risk averse) and the degree of uncertainties as measured by some statistical parameters. The mathematical complexities increase as more economic aspects are taken into consideration. Basically, this is a bounded state control problem in which the steady-state solution no longer has the significance of a dynamic equilibrium; rather, it yields the final levels of accumulated extractions and determines whether the resource will be depleted.

In Liu's paper, the problem of exploitation of exhaustible resources is tackled from a mathematical point of view. Some techniques for obtaining the optimum extraction rate for a firm over an infinite horizon are discussed. A general profit function and cost function associated with extraction are assumed. Uncertainty in the amount of reserve is also considered. He shows that at a fixed amount of accumulated extraction, the optimum extraction rate is always lower with uncertainty than without uncertainty. A significant feature of the analysis is the use of the transversality condition at infinity, from which the final level of accumulated extraction can be determined. Moreover, for n-firm exploitation, the transversality condition leads to an equilibrium condition among n firms under Pareto optimality. Different levels of accumulated extraction for different firms can also be obtained.

Arrow and Chang employ a model that includes both exploration and consumption of uncertain natural resources. The resource is assumed to be Poisson distributed throughout a relevant area. The amount of (unknown) reserve and the unexplored area at any time are the state variables, while the rate of consumption and the exploratory effort are the control variables. The economy seeks to maximize the integral of a utility function minus the cost of exploration. This is an optimal control problem with a jump process. Using the method of dynamic programming, they derive the equivalent of Bellman's equation. From that equation, the optimal policy and the optimum return function are characterized. The optimum exploratory effort almost always alternates between zero and infinity, and the impact of such an alternation is examined.

Lewis and Schmalensee investigate the implication of two intermediate market structures for supplier behavior. The first is a market dominated by a cartel that maximizes its profit subject to the price-taking behavior of a competitive fringe of many suppliers. An interesting result of their analysis is that when allowed to misrepresent the true extent of its reserve holdings, the cartel's optimal policy is not to lie but to tell the truth.

Lewis and Schmalensee also present a model of a Cournot–Nash oligopoly in which each firm is large enough to have some control over price. Existence and uniqueness of an equilibrium are established, and some comparative dynamic results are obtained under conditions of perfect information and homogeneous reserve holdings and extraction costs.

It has been long recognized in resource economics that risk taking plays an important role in a competitive market and affects resource allocation. Sutinen examines the implications of royalties for production over time. He notes that producers of petroleum and other exhaustible resources typically do not own the resource but rather purchase a lease giving them the right to extract the resource. A common feature of such leases is payment to the owner a royalty plus a fixed amount of payment.

Sutinen constructs a competitive model of the market for extractive rights, with resource owners being the suppliers and the producing firm the demander of the rights. Demand by consumers of the final product is assumed stochastic in all periods. Under these conditions, two sets of leasing arrangements are analyzed: a fixed rental and a class of risk-sharing arrangements involving royalty payments. The resulting sequence of production is characterized for each set of leasing arrangements and compared with the optimal production sequence. When both owners and firms are risk averse, Sutinen finds that royalties yield a production sequence superior to that of a fixed rental lease.

6

Optimal Pricing, Use, and Exploration of Uncertain Natural Resource Stocks†

KENNETH J. ARROW AND SHELDON S. L. CHANG

1. Introduction

The classic Hotelling (Ref. 1) model of exploitation of exhaustible resources assumes in its simplest form that the stock of the resource is known from the beginning. If there are no extraction costs, then the shadow prices associated with an optimal extraction policy rise at the rate of the market rate of interest. The only variable that has to be determined is the initial price, which then determines all future prices; the initial price depends on the interaction of demand (or utility) considerations with the initial stock. In a competitive world, prices would clearly have to rise at the rate of interest to keep resource holders indifferent between extracting the resource now and later.

If there are extraction costs, then the market price is the sum of the marginal extraction cost and the rent on the scarce stock. The latter still obeys the rule of the preceding paragraph, so that the rent will rise at the rate of interest.

The experience of many minerals, most strikingly oil from 1950 to 1970, is that the theoretically derived increase in prices is not observed.

† A preliminary version of this paper was originally presented by Arrow at the Conference on Natural Resource Pricing. Trail Lake, Wyoming, 15–17 August 1977 and later at the Third Kingston Conference on Differential Games and Control Theory, 5–8 June 1978. Chang was present at the latter conference and subsequently made the main contributions to completing the analysis.

KENNETH J. ARROW • Department of Economics, Harvard University, Cambridge, Massachusetts. SHELDON S. L. CHANG • Department of Electrical Sciences, State University of New York, Stony Brook, New York.

One might take a very long run view of the world and say that the post-1970 increases are making up for lost time. Perhaps the Hotelling theory looks better when 1977 is compared with 1950. Still such an evasion gives the theory little value.

No doubt many explanations for the discrepancy are possible. But certainly one rather obvious one is that the stock of the resource is far from known. Hence, new discoveries yielding upward revision of estimated reserves change the basis of calculation for the Hotelling rents. The predicted rise at the rate of interest is offset by repeated downward revision of the initial price in response to changing estimates of supply.

But if the stocks are in fact uncertain, as evidenced by repeated changes in estimates, then that uncertainty should be reflected in the initial planning. It is the aim of this paper to begin such an analysis.†

It is important, in our judgment, to distinguish between the process of learning about reserves and the consumption of them.‡ The first constitutes exploratory activity. It is clearly not the case that the only form of information about resource reserves comes from pumping for consumption.

The analysis in this paper is largely heuristic, and some points remain to be clarified, as will be seen.

2. The Model of Uncertainty about Resources

Resources are assumed to be distributed randomly throughout the relevant area; they can be discovered by costly exploration.

In particular, it will be assumed that there is no spatial correlation of resource distribution. In probability terms, the resource quantities in two

† The analysis of extraction and consumption policy under conditions of uncertainty about the amount of reserves has been studied a number of times in the literature. For a survey see Crabbe (Ref. 2) and for a recent example see Loury (Ref. 3).

‡ Pindyck (Ref. 4) has distinguished between exploration and extraction but not in the context of uncertainty. Gilbert (Ref. 5) has introduced exploration as a means of gaining information, as in the present paper, but his probability assumptions are different. Like Loury (Ref. 3), he assumes a given probability distribution for the total stock; extraction and exploration change the distribution only in the sense that the posterior distribution is conditioned on the statement that the total is at least as great as the amount extracted and explored for. This assumption is not grounded on the chance distribution of resources over the surface of the earth.

The model of Deshmukh and Pliska (Ref. 6), developed independently of the present paper, has fundamentally the same probability assumptions (strictly speaking, theirs are more general). However, they assume that there is no limit on unexplored land and hence in the long run the resource is not exhaustible.

overlapping areas are assumed to be independent random variables.† It turns out that this assumption, together with the natural condition that resource pools are nonnegative, severely restricts the range of possible distribution to be used. In particular, the resources must be located in discrete spots, randomly distributed over the area. If the mines are equally likely to be found in all parts of the area, then two natural assumptions are that the probability of finding a mine in an area h is, for small h, proportional to h, while the probability of finding two or more mines in a small area goes to zero more rapidly than h. In symbols,

$$\lim_{h\to 0} \frac{1}{h} \text{Prob (one mine in area } h) = \lambda,$$

$$\lim_{h\to 0} \frac{1}{h} \text{Prob (two or more mines in area } h) = 0,$$

where λ is some positive constant.

As is well known [see, e.g., Karlin and Taylor (Ref. 8), pp. 22–26], these assumptions imply a Poisson distribution of mines. Specifically, in an area A,

$$\text{Prob } (m \text{ mines in area } A) = \frac{e^{-\lambda A}(\lambda A)^m}{m!} \qquad (m = 0, \ldots).$$

This process shall be applied in the following way. At any time t, the *rate of exploratory effort* is the area explored per unit time. Let

$$x(t) = \text{rate of exploratory effort at time } t.$$

Then in the interval $(t, t + dt)$ for small dt, the area explored is $x(t)\,dt$. If we let $h = x(t)\,dt$, then we are assuming that

$$\text{Prob (discovering one mine between } t \text{ and } t + dt) = \lambda x(t)\,dt. \qquad (1)$$

The probability of discovering two or more mines in this small interval is infinitesimal compared with (1) and can be disregarded.

The learning in this model is simple but not nonexistent. It is true that exploration in any area gives no information about resources outside that area. However, the unexplored area is diminishing; the reserves are the sum of those known to be in the explored area plus a random variable representing reserves in the unexplored area, but the variance of the latter component is diminishing as the unexplored area diminishes. Let

$$X(t) = \text{area unexplored at time } t.$$

† Some support for this view can be found in the model of Menard and Sharman (Ref. 7, p. 337).

By definition,

$$\dot{X}(t) = -x(t), \tag{2}$$

where the dot denotes differentiation with respect to time, $X(0)$ is given, and $X(t)$ is restrained to be nonnegative; presumably, $X(t)$ approaches zero as t increases.

We make now one central simplifying assumption: all mines have the same resource quantities. This could be generalized by assuming that there is a distribution of mine sizes, for example, logarithmic normal. However, we believe that this generalization can be accommodated with little change in the following analysis. Without loss of generality, the amount of resource in each mine will be taken as one.

Let

$R(t) =$ known reserves at time t,

$N(t) =$ number of mines discovered between 0 and t,

$c(t) =$ rate of consumption of the resource at time t.

The total consumption of the resource up to time t is then

$$\int_0^t c(t)\,dt.$$

The known reserves at time t are the reserves at time 0 less consumption up to time t plus the number of mines discovered between 0 and t:

$$R(t) = R(0) - \int_0^t c(t)\,dt + N(t). \tag{3}$$

The variable $N(t)$ is a random variable; it increases by one unit in any small interval $(t, t + dt)$ with a probability given by (1). Therefore $R(t)$ is also a random variable. From (3),

$$\dot{R}(t) = -c(t) \qquad \text{if } t \text{ is between jump points of } N(t). \tag{4}$$

Known reserves $R(t)$ are constrained to be nonnegative everywhere.

3. The Criterion Function

The maximand will be the usual integral of discounted utilities. Here, utility at any point of time depends on both the amount of the resource being discussed and on goods in general. Since even the most important mineral requires a relatively small fraction of national income for production, it is reasonable to assume the absence of income effects, so that utility is linear in goods in general. It could therefore be ignored in analysis except that we will regard exploration as a costly activity. The inputs to exploration will be assumed to be goods in general. Let

$$P = \text{price of exploration in terms of goods in general,}$$

$$U(c) = \text{utility of consumption of resource at any instant.}$$

The units of measurement of U can be taken to be goods in general; hence, the utility generated at any instant is $U(c) - Px$, when the consumption is c and the rate of exploration is x.

The payoff to the economy is

$$\int_0^{\infty} e^{-\rho t}[U(c(t)) - Px(t)]\, dt, \tag{5}$$

where ρ = rate of time discount of utilities.

However, since constraint (3) involves a random variable, the payoff (5) will be a random variable, because future consumption and exploration will depend on the random discoveries made in the intervening period. The economy will therefore seek to maximize

$$E\left\{\int_0^{\infty} e^{-\rho t}[U(c(t)) - Px(t)]\, dt\right\}. \tag{6}$$

Criterion (6) is to be maximized with respect to the instruments $c(t)$, $x(t)$ subject to the constraints that $X(t) \geq 0$, $R(t) \geq 0$ everywhere, with $X(t)$ and $R(t)$ being defined by (2) and (3) respectively.

4. Analysis of the Optimal Policy

The methods are those of dynamic programming; see Bellman and Dreyfus (Ref. 9) for a heuristic exposition.

We consider the maximum value of (6) (under constraints) as a function of the initial values of the two *state variables* R and X. That is, we let

$$V(R, X) = \max E\left\{\int_0^{\infty} e^{-\rho t}[U(c(t)) - Px(t)]\, dt\right\} \quad \text{when } R(0) = R, X(0) = X. \tag{7}$$

Because time enters explicitly only through the exponential discount rate, the optimum starting from $t_0 > 0$ and discounted back to t_0 (rather than zero) would be the same function of $R(t_0)$, $X(t_0)$ that is given by (7). Hence, $V(R, X)$ is the sum of the integral over the interval $(0, t_0)$ and the expected value of V at time t_0, discounted back to time 0:

$$V(R, X) = \int_0^{t_0} e^{-\rho t}[U(c(t)) - Px(t)]\,dt + e^{-\rho t_0}E\{V(R(t_0), X(t_0))\}. \quad (8)$$

Note that $R(t_0)$ is a random variable, because its value depends on the number of discoveries in the interval $(0, t_0)$.

Equation (8) is valid if $c(t)$ and $x(t)$ have been chosen optimally. Alternatively, we may consider $c(t)$ and $x(t)$ to be variables over the interval $(0, t_0)$ and optimize over them, on the assumption that $V(R(t_0), X(t_0))$ is itself calculated on the assumption that an optimal policy is followed thereafter. Notice that $R(t_0)$ and $X(t_0)$ both depend on the choice of instruments over $(0, t_0)$, and hence the optimization must take account of the effects on both terms.

Now assume that $t_0 = dt$, a number sufficiently small that c and x may be regarded as constant over $(0, dt)$. We can also disregard the discounting within this interval. Hence, the first term in (8) can be approximated by

$$[U(c) - Px]\,dt.$$

The factor $e^{-\rho\,dt}$ is approximately $1 - \rho\,dt$. From (3), $R(dt)$ is a random variable. With c treated as constant,

$$R(dt) \approx R(0) - c\,dt + N(dt),$$

where the symbol $\approx$ means approximately equal. But in the small interval $(0, dt)$, the probability of discovering more than one mine is negligible. From (1), $N(dt) = 1$ with probability approximately $\lambda x\,dt$, and $N(dt) = 0$ with probability $1 - \lambda x\,dt$. Also, $X(dt) \approx X(0) - x\,dt$. Therefore,

$$E\{V(R(dt), X(dt))\} \approx (1 - \lambda x\,dt)V(R(0) - c\,dt, X(0) - x\,dt) + \lambda x\,dt\,V(R(0) + 1 - c\,dt, X(0) - x\,dt). \quad (9)$$

Assume in addition that V is a differentiable function of R and X. Let $V_R^0 = \partial V/\partial R$ evaluated at $(R(0), X(0))$, $V_X^0 = \partial V/\partial X$ evaluated at the same point, and let V_R^1 and V_X^1 be the two partial derivatives evaluated at $(R(0) + 1, X(0))$. Also note that $R(0) = R$, $X(0) = X$.

$$V(R(0) - c\,dt, X(0) - x\,dt) \approx V(R, X) - V_R^0 c\,dt - V_X^0 x\,dt,$$
$$V(R(0) + 1 - c\,dt, X(0) - x\,dt) \approx V(R + 1, X) - V_R^1 c\,dt - V_x^1 x\,dt.$$

Substitution into (9), some rearrangement, and discarding terms in $(dt)^2$ yield,

$$E\{V(R(dt), X(dt))\} \approx V(R, X) - V_R^0 c\,dt - V_X^0 x\,dt + \lambda x\,dt\,\Delta V(R, X).$$

(The operator Δ means the difference between the function at $R + 1$ and at R, X being held constant.)

If we substitute this and the other results into (8), subtract $(1 - \rho\, dt)$ $V(R, X)$ from both sides, divide through by dt, and let dt approach zero, we find that

$$\rho V(R, X) = \max_{c,x} \{U(c) - Px - V_R^0 c - V_X^0 x + \lambda x\, \Delta V\}. \tag{10}$$

Equation (10) will not hold throughout the (R, X)-plane. Note that no upper bound has been set on the exploration rate and that exploration costs are linear. Therefore, it is possible and, as will be seen shortly, true that there will be moments of infinitely rapid exploration. We may think of a whole area's being explored instantaneously, with a corresponding cost that is finite in total but incurred at an infinite rate. In such a case, there is an instantaneous downward jump in the state variable X (unexplored land). Equation (10) will not hold in the region of infinitely rapid exploration.

Now maximize (10) with respect to c and x, the instrument values at time 0. For the consumption rate c, let us ignore the corner possibility $c = 0$; this is certainly legitimate if $U'(0) = +\infty$, i.e., if the resource is indispensable. We then have

$$U'(c) = V_R^0. \tag{11}$$

The maximand in (10) is linear in x. One cannot exclude the possibility of either a corner solution, $x = 0$, or of an infinitely rapid exploration; indeed, one of these two cases will hold almost everywhere. Clearly, from (10), the rate of exploration x would be infinite if

$$P + V_X^0 < \lambda\, \Delta V$$

and would be zero if

$$P + V_X^0 > \lambda\, \Delta V.$$

Clearly, V_R^0 and V_X^0 are the shadow rents of the stock of reserves and the stock of unexplored land, respectively. We can drop the superscript 0, since the analysis holds at any time t. Let

$$p_R = \text{rent of reserves} = V_R,$$

$$p_X = \text{rent of unexplored land} = V_X.$$

Then (11) can be rewritten as

$$U'(c) = p_R. \tag{11'}$$

Let

$$A_\infty = \{(R, X) \mid P + p_X < \lambda\, \Delta V\}, \tag{12}$$

$$A_0 = \{(R, X) \mid P + p_X > \lambda\, \Delta V\}, \tag{13}$$

$$B = \{(R, X) \mid P + p_X = \lambda\, \Delta V\}. \tag{14}$$

Then, from earlier remarks,

$$\text{exploration is infinitely rapid for } (R, X) \in A_\infty, \tag{12$'$}$$

$$\dot{x} = 0 \quad \text{for } (R, X) \in A_0. \tag{13$'$}$$

When $(R, X) \in B$, it is so far possible that there will be exploration at a positive but finite rate; in any case, B is the boundary between the two regions A_0 and A_∞.

Equation (11′) is familiar; the shadow price of consumption must equal the shadow price of holding reserves. Relations (12′)–(13′) are more novel. The left-hand sides in the definitions (12)–(14) represent the costs of exploration, direct costs, and the use of unexplored land, which has a value due to its potential for resources. The right-hand sides represent the benefits, the gain in total utility (a finite increment, not a rate of change) due to finding a mine multiplied by the probability of finding one per unit effort.

We first investigate the structure of the optimal policy and of the return function V in region A_0. Here $x = 0$, c is given by (11′), and

$$\rho V = U(c) - V_R c,$$

from (10). Differentiate with respect to R and with respect to X; since c maximizes the right-hand side, it can be regarded as constant in accordance with the "envelope theorem":

$$\rho V_R = -V_{RR} c, \qquad \rho V_X = -V_{RX} c. \tag{15}$$

Divide the first equation by the second.

$$V_R / V_X = V_{RR} / V_{RX}.$$

This partial differential equation holds throughout the domain A_0. As is well known, it is equivalent to the statement that there exist functions $\bar{V}$ and R_E such that,

$$V(R, X) = \bar{V}[R + R_E(X)] \qquad \text{for } (R, X) \in A_0. \tag{16}$$

Without loss of generality, we may set

$$R_E(0) = 0. \tag{17}$$

Note that $R_E(X)$ may be regarded as the *resource-equivalent* value of unexplored lands.

The importance of this result is that the function $\bar{V}$ is the return function for the optimal solution in the case where there are no unexplored lands, as can be seen by setting $X = 0$ in (16). This is a standard problem which is easily solved. Thus, in the case where $U(c) = \ln c$, it is easy to see that

$$\bar{V}(R) = (\ln \rho - 1 + \ln R)/\rho.$$

Then, in the domain A_0, the instrument variables c and x are determined by the relations $x = 0$ and, from (11′),

$$U'(c) = V_R = \bar{V}'[R + R_E(X)].$$

The movements of the shadow prices in A_0 can also be established easily. Note that

$$V_{RR} = \bar{V}''[R + R_E(X)].$$

Then

$$\frac{1}{p_R}\frac{dp_R}{dt} = \frac{1}{V_R}\frac{dV_R}{dt} = \frac{1}{V_R} V_{RR}\dot{R};$$

since there is no exploration, $\dot{R} = -c$, by (4). Then by the first equation in (15),

$$\dot{p}_R/p_R = \rho \qquad \text{for } (R, X) \in A_0, \tag{18}$$

as in the usual Hotelling theory. From (16),

$$p_X = V_X = \bar{V}'R_E'(X); \tag{19}$$

hence the ratio p_X/p_R depends only on X and is therefore constant along any given path. Hence, we must also have,

$$\dot{p}_X/p_X = \rho \qquad \text{for } (R, X) \in A_0. \tag{20}$$

This regime continues, with R decreasing, until the trajectory hits B. Note that the behavior has been defined up to still unknown function R_E.

Now let us consider the determination of the return function $V(R, X)$ in A_∞. Suppose the initial situation is $(R, X + dX)$. Exploration occurs at an infinite rate, so consumption can be disregarded. When the area dX is explored, the probability of a discovery is $\lambda\, dX$ to a first approximation. If a discovery is made, the point $(R, X + dX)$ is transformed into $(R + 1, X)$; if not, it is transformed into (R, X). The cost incurred is $P\, dX$. Hence,

$$V(R, X + dX) \approx (\lambda\, dX)V(R + 1, X) + (1 - \lambda\, dX)V(R, X) - P\, dX.$$

Dividing through by dX and letting dX approach 0 yields

$$V_X(R, X) = \lambda\, \Delta V(R, X) - P \qquad \text{for } (R, X) \in A_\infty. \tag{21}$$

Let the boundary B be represented by the curve $R = R_B(X)$, so that

$$R > R_B(X) \quad \text{for } (R, X) \in A_0, \qquad R < R_B(X) \quad \text{for } (R, X) \in A_\infty. \tag{22}$$

We wish to determine $R_B(X)$. Since the point $(R_B(X), X)$ is on the boundary of both A_0 and A_∞, it follows by continuity that V_X is given by both (19) and (21). Thus

$$\bar{V}'[R_B(X) + R_E(X)]R_E'(X) = \lambda \, \Delta V[R_B(X), X] - P. \tag{23}$$

For a path starting from some (R, X), $R > R_B(X)$, $V(R, X) = \bar{V}[R + R_E(X)]$ is maximized when $R_E(X)$ is maximized. If R_E is given up to and including X, it follows that we wish to maximize $R_E'(X)$, since this act maximizes $R_E(X + dX)$. From (23), then,

$$R_E'(X) = \max_r \, [\lambda \, \Delta V(r, x) - P]/\bar{V}'[r + R_E(X)], \tag{24}$$

$R_B(X)$ is the value of r that achieves the maximum in (24). (25)

These equations, together with (21) and (16), completely describe the solution. Start with $V(R, 0) = \bar{V}(R)$, a known function. Having defined the functions $V(R, X)$ and $R_E(X)$ up to some point X for all R, we can calculate $R_B(X)$ and $R_E'(X)$ from (24) and (25). We can then define $V(R, X + dX)$ from (21) for $R < R_B(X)$, $R_E(X + dX) = R_E(X) + R_E'(X)\, dX$, and therefore $V(R, X + dX) = \bar{V}[R + R_E(X + dX)]$ for $R > R_B(X)$.

One last important point. It can be shown that R_B is a decreasing function. Suppose, to the contrary, that it were increasing in some interval. Start with a point $(R, X + dX) \in B$, so that $R = R_B(X + dX)$. If exploration over dX produces no discovery, then we arrive at the point (R, X); but $R = R_B(X + dX) > R_B(X)$, so that $(R, X) \in A_0$. If a discovery is made, then we arrive at $(R + 1, X)$, which certainly belongs to A_0. In that case,

$$\Delta V(R, X) = \bar{V}[R + 1 + R_E(X)] - \bar{V}[R + R_E(X)].$$

In the maximization in (24), $R_E(X)$ is taken as given. Hence, maximizing over r is equivalent to maximizing over $s = r + R_E(X)$. Therefore, (24) becomes

$$R_E'(X) = \max_s \, [\lambda \, \Delta \bar{V}(s) - P]/\bar{V}'(s);$$

and

$$R_B(x) = s^* - R_E(X),$$

where s^* achieves the maximization of $R_E'(X)$. But the maximand is now independent of X, so s^* is independent of X. Since $R_E(X)$ is certainly increasing (an increase in unexplored lands can certainly never decrease total return), $R_B(X)$ must be decreasing, a contradiction to the original assumption that R_B is increasing.

That R_B is decreasing is a bit surprising at first. However, suppose that there is an infinite amount of unexplored territory, the case studied by Deshmukh and Pliska (Ref. 6). Under our special assumptions (linear costs of exploration), it is easy to see that the optimal path starting with some reserves R is to exhaust them along an optimal path (with prices rising at the rate ρ). When $R = 0$, then exploration takes place at an infinite rate until a discovery is made. There is no reason to explore earlier; there is no information gained, since a discovery will always be made with probability one and, in view of discounting, it pays to postpone exploration until the last possible moment. When the discovery is made, exploration ceases, and consumption again proceeds until the new reserves are exhausted.

In this case, the price rises at the rate ρ until exhaustion but drops abruptly on discovery. Over a long period of time, price has no trend, though it varies periodically.

We may assume that in our model, with large X, the effect is approximately the same as infinite X. Clearly, with $U'(0) = +\infty$, one would not proceed to absolute exhaustion; with a finite X, there is always a finite probability that there are no discoveries to be made. But with large X, the probability of that event, $e^{-\lambda X}$, is very small, so $R_B(X)$ would be expected to be very small. On the contrary, with small X, information about the remaining reserves becomes more important, so that exploration takes place at a higher level of reserves.

The price history will show fluctuations with little upward trend when X is large; presumably the upward trend is stronger as X approaches zero, but this requires a probabilistic analysis not yet performed.

5. Commentary

The alternation of zero and infinite rates of exploration is an unfelicitious feature of the model. It is due to the assumption that exploration costs are flows and can be turned on and off without cost. If one added to the model the need for capital invested in exploration and production, there would be a tendency to smooth out exploration activities in order to make better use of the capital. But this would convert the problem into one with three state variables, which would be even more difficult.

References

1. Hotelling, H., The economics of exhaustible resources, *Journal of Political Economy*, Vol. 39, pp. 137–175, 1931.
2. Crabbe, P., L'Exploration des ressources extractives non renouvelables: Théorie économique, processus stochastique et vérification, *L'Actualité Économique*, Montreal, Canada, pp. 559–586, October–November 1977.
3. Loury, G. C., *The Optimum Exploitation of an Unknown Reserve*, Review of Economic Studies, Vol. 45, pp. 621–636, 1978.
4. Pindyck, R. S., Optimal exploration and production of a nonrenewable resource, *Journal of Political Economy*, Vol. 86, pp. 841–862, 1978.
5. Gilbert, R. J., Optimal depletion of an uncertain stock, *Review of Economic Studies*, Vol. 45, pp. 47–58, 1979.
6. Deshmukh, S. D. and Pliska, S. R., *Optimal Consumption and Exploration of Nonrenewable Resources under Uncertainty*, Discussion Paper No. 317, Center for Mathematical Studies in Economics and Management Science, Northwestern University, 1978.
7. Menard, H. W. and Sharman, G., Scientific uses of random drilling models, *Science*, Vol. 190, pp. 337–343, 1975.
8. Karlin, S. and Taylor, H. M., *A First Course in Stochastic Processes*, 2d ed., Academic Press, New York, San Francisco, and London, 1975.
9. Bellman, R. and Dreyfus, S., *Applied Dynamic Programming*, Princeton University Press, Princeton, N.J., 1962.

7

Restricted Access to Common-Property Fishery Resources: A Game-Theoretic Analysis

COLIN W. CLARK

1. Introduction

The "tragedy of the commons" (Ref. 1) has proved particularly difficult to counteract in the case of marine fishery resources (Refs. 2–4), where the establishment of individual property rights is virtually out of the question. Common ownership is the fundamental fact affecting almost every regime of fishery management.

Property rights in marine fisheries have, however, undergone a revolutionary transformation over the past few years, resulting from the declaration by many coastal states of 200-mile zones of extended fisheries jurisdiction. While the fishery resources within such zones have thus become appropriated by the states in question, in most cases they still remain common property as far as individual users of the resource are concerned. An important philosophical question—with significant practical implications—is whether property rights to natural resources should rest with the state or be conferred on specific individuals or groups. But such questions are beyond the scope of this discussion.

Traditional methods of fishery management (Ref. 5) have concentrated primarily on the protection of fish stocks, generally by means of physical methods such as seasonal or area closures, total catch quotas, or

COLIN W. CLARK • Department of Mathematics, University of British Columbia, Vancouver, British Columbia, Canada.

restrictions on vessel or gear characteristics. Although such techniques have often proved highly successful for the purpose of biological conservation, they have provided few economic benefits to the fishing industry, or to society at large. Elementary economic analysis (Ref. 6) shows that, regardless of physical constraints, unrestricted access to fishery resources leads inevitably to the dissipation of economic rents. The development of almost every commercial fishery attests to the accuracy of this prediction.

As a result, an observable tendency toward restricting entry to fishery resources has developed. Examples include the Pacific salmon fisheries, Northwest Atlantic fisheries generally, and the crustacean fisheries of Australia. Many other fisheries appear to be on the verge of the establishment of limited-entry programs, as nations attempt to rationalize coastal resource exploitation.

It is unfortunate, therefore, that existing programs of this kind have not been notably successful from the economic point of view. For example, the British Columbia salmon license limitation program (initiated in 1968) has failed to control capital inputs or to generate significant long-run economic benefits. The reaction of vessel owners to the license program, in terms of capital intensification, seems not to have been anticipated, even though it could have been predicted from relatively simple theoretical considerations.

Most published analyses of fisheries economics (Refs. 2, 3, 6–11) have been concerned with two contrasting systems of property rights: full rights (sole ownership) and no rights (open access). Little study has been directed to the case of limited access, and it is this omission that the present paper attempts to rectify.

Models of fishery exploitation under sole ownership are particularly convenient analytically, since it can reasonably be hypothesized that the sole owner's objective is to maximize the present value of net economic revenues. The optimal harvest policy can be deduced by optimal control theory (Ref. 3). The situation is by no means so clear for the case of $N>1$ "owners," or producers. In this paper we consider the limited-access fishery as an N-person, nonzero-sum differential game, but the following warnings of the founder of differential games should be kept in mind: "Differential games are the same step from finite games that [optimal control theory] is from maximizing a finite set. ... the very meaning of a solution is far from clear" (Isaacs, Ref. 12).

In Section 2 we briefly review the theory of the sole owner; the results of this section are used extensively later. In Section 3 we discuss the case of N producers, with no restrictions on each producer's input levels. Section 4 introduces constraints on these inputs. In Section 5 the theory is applied to the study of various forms of fishery regulation.

The paper should be considered an introduction to the theory of limited-entry fisheries, since many important complications are not discussed. In particular, we study only a simple deterministic (and autonomous) model of the fishery; more complex models involve a much more difficult analysis.

2. The Basic Model

In this section we describe briefly a model of the commercial fishery due to Gordon (Ref. 6) and Schaefer (Ref. 13) that has been extensively employed in the economic literature (Refs. 8–10). The model equations are

$$\frac{dx}{dt} = G(x) - h(t), \qquad x(0) = x_0, \tag{1}$$

$$h(t) = qE(t)x(t), \tag{2}$$

$$x(t) \geq 0, \tag{3}$$

$$\pi(x(t), E(t)) = ph(t) - cE(t) = (pqx(t) - c)E(t), \tag{4}$$

$$PV[E] = \int_0^\infty e^{-\delta t}\pi(x(t), E(t))\, dt, \tag{5}$$

$$0 \leq E(t) \leq E_{\max}, \tag{6}$$

where the symbols have the following interpretations:

$x(t)$ = biomass of fish population at time t,
$G(x)$ = natural growth function,
$h(t)$ = rate of catch,
$E(t)$ = rate of fishing "effort",
q = catchability coefficient,
p = landed price of fish,
c = cost of effort,
δ = discount rate,
$\pi(x, E)$ = flow of economic rent,
$PV[E]$ = present value of economic rent,
$E_{\max}$ = maximum effort capacity.

The growth function and the various parameters are assumed to be known; $G(x)$ is assumed to satisfy

$$G(0) = G(\bar{x}) = 0; \qquad G(x) > 0 \qquad (0 < x < \bar{x}),$$

where $\bar{x}$ denotes "carrying capacity."

A full analysis of this model (and some of its variations) appears in Refs. 2 and 8; the details will not be repeated here. We shall, however, identify two particular equilibria that are often associated with the model; the *bionomic equilibrium* of the open-access fishery and the equilibrium corresponding to *maximum* (discounted) *economic yield*, i.e., maximum present value.

According to the theory of Gordon (Ref. 6), bionomic equilibrium in the open-access fishery is established at the biomass level $x = x_\infty$ at which economic rent becomes zero, i.e., from equation (4) at

$$x_\infty = c/pq. \tag{7}$$

The corresponding level of effort is

$$E_\infty = G(x_\infty)/qx_\infty. \tag{8}$$

Maximum economic yield, on the other hand, corresponds to an effort policy $E(t)$ that maximizes the present value $PV[E]$. This problem can be solved by an elementary argument (Ref. 14). Introduce the expression

$$Z(x) = \int_{x_\infty}^{x} (p - c/qu)\, du. \tag{9}$$

Integrating by parts, we then obtain

$$\begin{aligned} PV[E] &= \int_0^\infty e^{-\delta t}(p - c/qx)(G(x) - \dot{x})\, dt \\ &= \int_0^\infty e^{-\delta t} V(x)\, dt + \text{const}, \end{aligned} \tag{10}$$

where

$$V(x) = (p - c/qx)G(x) - \delta Z(x). \tag{11}$$

The problem requires maximization of (10) by means of a path $x = x(t)$ satisfying the constraints

$$G(x) - qE_{\max}x \le \dot{x} \le G(x). \tag{12}$$

We assume $E_{\max}$ large enough that all biomass levels $x \in [x_\infty, \bar{x}]$ are feasible equilibria.

If $V(x)$ has a unique maximum at $x = \hat{x} \in (x_\infty, \bar{x})$, and if $V(x)$ is monotone above and below $\hat{x}$, the solution is obvious: the biomass should be driven from x_0 to $\hat{x}$ by the most rapid possible approach:

$$E(t) = \begin{cases} E_{\max}, & \text{if } x(t) > x, \\ G(\hat{x})/q\hat{x}, & \text{if } x(t) = \hat{x}, \\ 0, & \text{if } x(t) < \hat{x}. \end{cases} \tag{13}$$

Note that this conclusion does not depend on any special smoothness assumptions; for example $G(x)$ may even possess discontinuities (provided that the equation (1) remains meaningful). This will be important later.

Even if $V(x)$ has a finite number of local extrema, it can easily be shown (Ref. 14) that the optimal policy must be of the form (13), where $\hat{x}$ denotes a local maximum, possibly depending on the initial biomass level x_0.

[The parameter E_{max} in this model is largely ad hoc and could in fact be taken to be infinite, thus giving rise to the possibility of impulse controls. More realistically, E_{max} can be taken to represent the level of capital invested in fishing capacity, in which case it becomes an additional decision variable (Ref. 15). This complication will not be pursued here, however.]

The relative simplicity of this model, which stems from its linearity in the control variable, allows us to deduce the optimal policy in feedback form (13). The game-theoretic model of the restricted-access fishery to be discussed in the next section can also be solved explicitly in the linear version. Linearity enters our model in three respects: in the production relation (2), in the assumptions of infinite elasticity of demand for fish, and infinite elasticity of supply of effort [equation (4)]. In practice, all three assumptions would be questionable. In particular, the assumption that (operating) cost is a linear function of effort, while it may be a reasonable approximation for a firm (or national industry) that can own and operate a fleet of any desired size, is obviously unrealistic for the decision facing an individual vessel owner.

The simplest reasonable model for the individual fishing vessel would employ a nonlinear function $E = f(L)$ for the "production" of fishing effort (Ref. 7), where L in general denotes a vector of inputs. Such nonlinear models will be discussed further in Section 4.

3. Competitive Equilibrium Involving N Producers

We now take up the study of the competitive limited-access fishery, first considering the case of N producers (N fixed), each having unrestricted rights to exploit the stock in competition with the other producers. This situation will be analyzed by means of the Nash theory of N-person (differential) games (Refs. 12, 16, 17).

The analysis presented in this section seems appropriate for studying the behavior of a fixed number of firms (or of nations) exploiting a common stock under competitive conditions. A more realistic model of the open-access fishery exploited by a number of individually owned vessels will be described in the next section.

We begin by examining the case $N = 2$; the general case will follow as a natural extension. With evident notation, our model now becomes

$$\frac{dx}{dt} = G(x) - h_1(t) - h_2(t), \qquad x(0) = x_0, \tag{14}$$

$$h_i(t) = q_i E_i(t) x(t), \tag{15}$$

$$x(t) \geq 0, \tag{16}$$

$$\pi_i(x, E_i) = (p_i q_i x - c_i)E_i, \tag{17}$$

$$PV_i[E_1, E_2] = \int_0^\infty e^{-\delta_i t}\pi_i(x(t), E_i(t))\,dt \tag{17a}$$

$$0 \leq E_i(t) \leq E_i^{\max}. \tag{18}$$

Thus in general the two producers may face different prices, costs, discount rates, and fishing technologies.

By definition, a (closed-loop) Nash competitive solution to this two-person, nonzero-sum differential game consists of a pair of control policies E_i^* in feedback form,

$$E_i^* = E_i^*(x),$$

satisfying the inequalities

$$PV_1[E_1^*, E_2^*] \geq PV_1[E_1, E_2^*] \qquad \text{for all } E_1, \tag{19}$$

$$PV_2[E_1^*, E_2^*] \geq PV_2[E_1^*, E_2] \qquad \text{for all } E_2. \tag{20}$$

Thus if producer 2 uses the Nash policy E_2^*, then producer 1 can do no better than use E_1^*, and vice versa.

The extent to which the Nash concept captures the actual motivation of competitive producers is subject to question. Even if explicit cooperation is assumed away (Ref. 18), it might be argued that some form of "silent negotiation" could lead to alternative equilibria (Ref. 19). (The latter possibility seems especially likely if capital, as represented by $E_i^{\max}$, is assumed costly and nonmalleable, but this assumption introduces many new difficulties.) On the other hand, it might be argued (Ref. 16) that the "maximin" solution is more realistic:

$$PV_1[E_1^*, E_2^*] = \max_{E_1} \min_{E_2} PV[E_1, E_2].$$

In our model, this concept assumes that producer 2 will operate at a loss merely for the discomfiture of producer 1, which seems rather unlikely.

Let x_i^* denote producer i's sole-owner optimal biomass level, as determined in Section 2; we now assume that x_i^* is uniquely determined, independent of the starting level x_0. Also let $x_i^\infty = c_i/p_i q_i$. We will say that producer 1 is more "efficient" than producer 2 if

$$x_1^\infty < x_2^\infty.$$

Theorem 3.1. Suppose that producer 1 is more efficient than producer 2. Then provided that $E_1^{\max}$ and $E_2^{\max}$ are sufficiently large, the following equations specify a Nash competitive solution:

$$E_1^*(x) = \begin{cases} E_1^{\max}, & x > \min(x_1^*, x_2^\infty), \\ G(x)/q_1x, & x = \min(x_1^*, x_2^\infty), \\ 0, & x < \min(x_1^*, x_2^\infty); \end{cases} \quad (21)$$

$$E_2^*(x) = \begin{cases} E_2^{\max}, & x > x_2^\infty, \\ 0, & x \le x_2^\infty. \end{cases} \quad (22)$$

This result is not unexpected: by reducing the stock to an appropriate level, the more efficient agent eliminates his competitor. If his efficiency advantage is large, he may actually operate at his sole-owner optimum x_1^*; this is clearly a Pareto optimum. But if the efficiency advantage is less pronounced, the equilibrium will be at a suboptimal stock level. In the limiting case of equal efficiency of the two producers, economic rent will be completely dissipated at equilibrium. (In the latter case the distribution of effort between the two producers is clearly indeterminate.)

The proof of the theorem is by direct verification, using the results of Section 2. To prove (19), for example, define

$$G_1(x) = \begin{cases} G(x) - q_2E_2^{\max}x, & x > x_2^\infty, \\ G(x), & x \le x_2^\infty. \end{cases} \quad (23)$$

If $E_2^{\max}$ is large, we have $G_1(x) < 0$ for $x > x_2^\infty$. From (11) it is clear that producer 1's optimal biomass level [relative to $G_1(x)$] equals $\min(x_1^*, x_2^\infty)$. The corresponding optimal effort policy is given by (21), and this establishes (19). A similar argument proves (20).

This Nash solution is easily seen to be unique. Let $\bar{x}$ denote a biological equilibrium corresponding to given Nash policies $E_i^*(x)$, i.e., $G(\bar{x}) = [E_1^*(\bar{x}) + E_2^*(\bar{x})]q\bar{x}$. Clearly both policies $E_i^*(x)$ must be zero for $x < \bar{x}$. But if $\bar{x} > x_2^\infty$, such policies cannot constitute a Nash solution, since either producer can increase his *PV* by reducing x to $\bar{x} - \varepsilon$. A similar argument establishes uniqueness of the Nash solution at x_2^∞, with one exception: in the case that $x_1^\infty = x_2^\infty$, the distribution of effort at equilibrium is clearly indeterminate (and immaterial).

(We note in passing that these arguments—and the description of the Nash solution—require modification when either $E_i^{\max}$ is small. Details are omitted.)

The foregoing model assumes that both producers are present at the outset. Alternatively, suppose that producer 1 originally has sole possession of the fishery but anticipates additional entry. Then producer 1 will switch from his own optimal sustained yield at x_1^* to a more intensive exploitation policy *prior* to the anticipated additional entry.

The extension of these results to the case of N competitive producers is now straightforward. The fishery will see the progressive elimination of inefficient producers as the fish stock is reduced. Ultimately only the most efficient producers will survive; if two or more maximally efficient producers exist, the fishery will converge to the classical bionomic equilibrium, with zero sustained economic rent. Marginally efficient producers will always be on the verge of re-entry, however.

It would be interesting to compare the competitive solution with Pareto and social optima. Except in the case of equivalent producers, however, these concepts have various interpretations (e.g., are side payments permissible?). For the case of equivalent producers, the social optimum is evident: converge to the optimal biomass level x^* as rapidly as possible. The equilibrium distribution of effort $\sum_i E_i = E^*$ is immaterial. We conclude that, as expected, the competitive fishery ($N \geq 2$) results in overfishing (unless total capacity happens to be small).

The model described in this section can be modified to address the important question of overcapitalization. Specifically, assume that the parameter $E_i^{\max}$ is proportional to total capital owned by producer 1. This capital has a cost π_i and furthermore is nonmalleable in the sense that it cannot be transferred to other uses. What levels of capital $E_i^{\max}$ will each producer require?

Even for the case of a sole producer, the solution to this problem is nontrivial (Ref. 15). For $N > 1$ the (Nash) solution is not known; the capital levels must be handled as additional state variables. If $\tilde{x}_i$ denotes the total-cost rent-dissipation level, $\tilde{x}_i = (c_i + \delta_i\pi_i)/(p_i q_i)$, it might be conjectured that a Nash equilibrium would yield $x \leq \min \tilde{x}_i$. Preliminary study suggests this is not correct, however, although it is probably true in the limit as $N \to \infty$.

4. Restricting the Factors of Production

According to the previous section, a Nash competitive solution involving as few as two producers results in the long-run dissipation of economic rent. This conclusion depends, however, on an assumption that producers face an infinitely elastic supply of all factors of production relative to the application of fishing effort. But typical methods of licensing entrants to regulated fisheries make at least some attempt also to control the factors of production. Normally a licensee will be permitted to operate a single vessel, or a single vessel of a specific type or tonnage.

In theory it should be possible to control both the number of vessels and the maximum level of effort per vessel, to ensure that significant rents are generated. If significant rents are generated, each vessel owner then has a strong incentive to try to capture an additional share of those rents. Thus if any factors of production remain uncontrolled, or only partially controlled, the vessel owner will be tempted to purchase additional units of these factors. In this section we attempt to generalize our basic model to take account of such behavior.

For simplicity we shall consider a single-variable factor of production L. We also hypothesize a fixed production function relating L to "fishing mortality" f:

$$f = f(L); \tag{24}$$

by definition fishing mortality (represented in Section 3 by qE) is the proportional rate of mortality imposed on the fish stock by the fishing vessel under discussion. Without loss of generality we may suppose that the supply of L is itself infinitely elastic.

The equations of the model are the same as (14)–(18) except that (15) now becomes

$$h_i = f_i(L_i)x, \tag{25}$$

and (17) becomes

$$\pi_i(x, L_i) = p_i f(L_i)x - c_i L_i. \tag{26}$$

The constraint $L_i \le L_i^{\max}$ is dropped entirely. We shall consider only equivalent producers, so that subscripts i can usually be omitted. [Note that (14) and (25) imply that total fishing mortality is additive: $f_{\text{total}} = \sum_j f_j$. Thus our model ignores "crowding" externalities (Ref. 20)]

The model of Section 3 is of course a special case of the preceding one, obtained by specifying

$$f(L) = \begin{cases} qL, & 0 \le L \le L^{\max}, \\ qL^{\max}, & L > L^{\max}. \end{cases} \tag{27}$$

For the general case we assume that f is concave:

$$f(0) = 0; \qquad f'(L) > 0; \qquad f''(L) < 0. \tag{28}$$

Unfortunately I do not know how to solve this differential game. If

$$H_i = e^{-\delta_i t}[p_i f_i(L_i)x - c_i L_i] + \lambda_i\left[G(x) - \sum_{j=1}^{N} f(L_j)x\right], \tag{29}$$

we have the following necessary conditions for a Nash closed-loop solution (Ref. 16):

$$\frac{\partial H_i}{\partial L_i} = 0; \qquad \frac{d\lambda_i}{dt} = -\frac{\partial H_i}{\partial x} - \sum_{j \ne i} \frac{\partial H_i}{\partial L_j}\frac{\partial L_j}{\partial x}. \tag{30}$$

The presence of the last sum apparently renders these conditions useless for computation (Ref. 16). The first equation implies that

$$p_i - c_i/f_i'(L_i)x = \lambda_i e^{\delta_i t}. \qquad (31)$$

If producer i ignores the effect of harvesting on the biomass and merely maximizes current revenue π_i, he sets

$$f_i'(L_i) = c_i/p_i x. \qquad (32)$$

If we assume that $\lambda_i > 0$, then (31) implies that the Nash solution involves a less intensive rate of exploitation than this.

Actually, inasmuch as the present model is intended to portray the behavior of many individual fishermen, the assumption (32) is probably quite realistic, *except for the possibility of changing the production function.* Under this assumption we have simply, for the case of N equivalent fishermen,

$$\frac{dx}{dt} = G(x) - Nf(L)x, \qquad (33)$$

$$f'(L) = c/px. \qquad (34)$$

The solution is depicted in Fig. 1(a) for the case of equation (27) and in Fig. 1(b) for the general concave case. For fixed N, equilibrium is established at point Q. As N increases, the biomass equilibrium curve ($\dot{x} = 0$) moves to a lower position, and Q approaches the open-access bionomic equilibrium at $(x_\infty, 0)$.

Thus if we assume that the individual production function $f(L)$ is fixed irrevocably, we can conclude that entry limitation is an appropriate instrument for generating economic rent in the fishery. The more interesting practical question is whether the same result can be achieved by

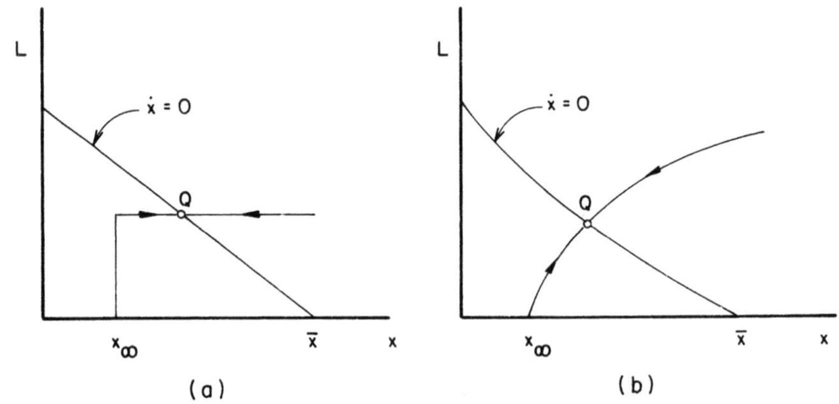

Figure 1. Solutions of equations (33) and (34).

controlling only some of the factors of production. In the Alaskan salmon fishery, for example, licenses are issued to individual fishermen, but vessel and gear characteristics do not seem to be fully controlled (Ref. 21). The British Columbia salmon program, on the other hand, licenses vessels by tonnage, but other inputs are not controlled. Substantial increases of capital have flowed into the latter fishery since the license program was established. Similar behavior has been reported in other limited-entry fisheries.

It may not prove possible to develop a theory that will convincingly answer this question. (Empirical analysis may prove equally difficult.) It seems that one must either make an *a priori* behavioral assumption that resolves the problem, or else one must fall back on game theory. But even aside from the technical difficulties of solving differential games, it is not necessarily clear which concept of solution is most appropriate. The analysis of the previous section at least suggests that whenever positive rents exist, there will be an incentive to expand whatever inputs remain uncontrolled. But perhaps this observation was evident from the outset.

5. Methods of Fishery Regulation

We now use our game-theoretic framework (such as it is) to analyze the effects of various methods for the regulation of common-property fisheries. The methods we shall consider are (1) vessel licenses, including fees; (2) taxes and royalties; (3) "physical controls," such as gear restrictions and seasonal closures; (4) quota systems.

Suppose that a license limitation program is initiated in a fishery originally at bionomic equilibrium. If the number of licenses issued is less than the original number of participants, then the licenses will acquire a positive value reflecting anticipated rents. Our analysis suggests, however, that these rents will likely be temporary, unless licensees are somehow prevented from increasing their capital (or other) inputs. Thus both theory and experience provide evidence that license limitation alone is unlikely to lead to rationalization in fishery management. (The possibility of a perpetual progressive license reduction scheme, which one sometimes hears recommended, does not seem worthy of serious thought.)

The method by which licenses are granted might influence their effectiveness, perhaps. For example, if licenses are sold at a price reflecting their capitalized value, it might be conjectured that the capital funds thereby "removed" from the fishery would inhibit subsequent expansion. But this does nothing to remove the incentive for such expansion. Unless capital markets are highly imperfect, one would expect the incentive to bear fruit.

Alternatively, license fees might be charged on an annual basis. Such fees could be adjusted periodically to reflect expected rents, and could also be related to estimated fishing power of vessels. In this case, the fees would be essentially equivalent to taxes on fishing effort and could be quite effective for extracting rent from the fishery. Indeed, it seems likely that this will be the favored technique for collecting rent from coastal fishery resources exploited by foreign vessels. The method is outlawed, however, for American vessels in U.S. waters by the 1976 Fishery Conservation and Management Act (see Christy, Ref. 22).

Taxes (royalties) on landed catch have long been recommended as the simplest and most direct method of capturing rent from fishery resources. There is little doubt that landings taxes could be highly effective, since they clearly affect fishermen's incentives. (It is doubtful, however, that landings taxes by themselves would provide a sufficiently accurate instrument to guarantee biological conservation—although simplistic models might lead to this conclusion. The need for direct biological control of fisheries will generally persist, although economic rationalization should greatly reduce the potential for serious depletion.) Of course, both license fees and taxes have distributional implications that are obviously odious to fishermen. An interesting question is whether economic efficiency can be achieved without withdrawing the rent from the fishery.

Before taking up this question let us briefly discuss traditional biological control methods. These include total catch quotas, escapement regulation, gear restriction, area closures, and seasonal openings and closures. The economic implications of these methods can only be deduced in detail by modifying the biological model appropriately. Biological controls may often be necessary to prevent severe depletion of fish stocks, and this is obviously of considerable economic significance. But unless access is restricted, there is no reason to suppose that such methods will lead to economic efficiency over the long run—there are too many counterexamples. Our game-theoretic analysis suggests that limited access is also unlikely to result in long-run economic efficiency, even if combined with biological controls.

For example, let us consider the possibility of total catch quotas. If these are imposed periodically with period T (normally, one year), the model of Section 3 can be modified by adding the constraint

$$\sum_i Y_{ik} \leq Q_k \qquad (k \geq 0), \tag{35}$$

where

$$Y_{ik} = \int_{kT}^{(k+1)T} h_i(t)\, dt \tag{36}$$

and where Q_k denotes the total catch quota for period k. Specifically, suppose that (as in the yellowfin tuna fishery in the eastern tropical Pacific) the season is closed as soon as the annual quota has been taken. Then $h_j(t)$ can only be positive, for $kT \leq t < (k + 1)T$, if

$$\sum_i \int_{kT}^{t} h_i(s)\, ds < Q_k. \tag{37}$$

For fixed effort capacities $E_i^{\max}$, the Nash competitive solution is easily determined. For example, for two equivalent producers we have simply

$$E_i(t) = \begin{cases} E_i^{\max} & \text{if } x(t) > x^{\infty} \text{ and (37) holds,} \\ 0 & \text{otherwise.} \end{cases}$$

Clearly the quota constraint (37) can be used to prevent x from being run down to x^{∞}; stock protection is the purpose of quota controls. But does the combination of quotas and limited entry guarantee the generation of economic rent? With no entry limitation, it can be predicted that the fishery will become overcapitalized, tending toward an equilibrium with returns to capital equal to opportunity cost. This overcapitalization will normally be manifested by a greatly shortened fishing season.

For the case of N (fixed) producers, we face the same difficulty mentioned in Section 3 for the capitalization game. To simplify matters, however, we can assume that the quota constraint is in fact binding (why else would quotas be imposed?), so that each producer employs $E_i = E_i^{\max}$ until the total quota is caught. This simplification completely removes our problem from the realm of differential games; the only choice left to producers is their level of capitalization. This problem can be handled by the usual reaction-curve approach.

Figure 2(a) shows typical curves ϕ_i (obtained by numerical simulation) for two equivalent producers. The Nash solution is at point Q. For $N > 2$ equivalent producers we obtain the reaction surfaces

$$E_i^{\max} = \phi\left(\sum_{j \neq i} E_j^{\max}\right).$$

It is easily seen that the Nash solution E_N satisfies

$$NE_N \rightarrow E_1^{\infty} \qquad \text{as } N \rightarrow \infty,$$

where E_1^{∞} is the amount of total capacity that inhibits further entry (Fig. 2a). We conclude that a combined limited-entry, total-quota management regime (as in the British Columbia salmon fishery) could lead to the generation of positive economic rents, but at a strictly suboptimal level approaching zero as N increases.

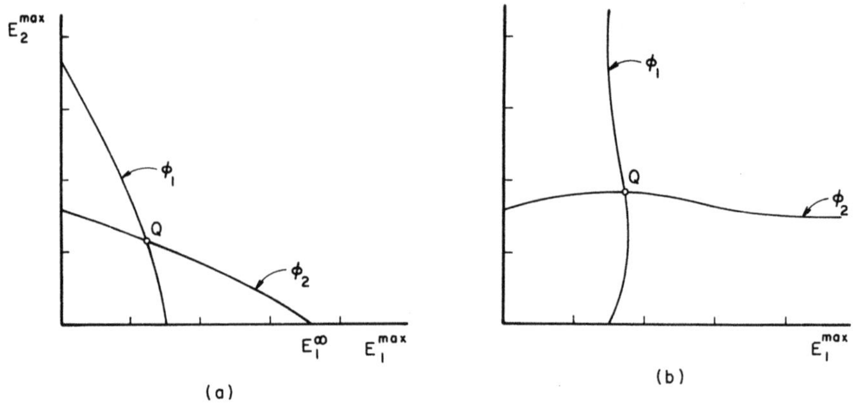

Figure 2. Reaction curves. (a) Total quota system. (b) Allocated quota system. The Nash equilibrium for two producers is at point Q.

The effect of a system of allocated catch quotas can be modeled by replacing equation (35) by

$$Y_{ik} \leq Q_{ik} \qquad (i=1,2,\ldots,N;\, k \geq 0), \tag{38}$$

where N represents the total number of quotas to be allocated. Then $h_j(t)$ can be positive, for $kT \leq t < (k+1)T$, only if

$$\int_{kT}^{t} h_j(s)\, ds < Q_{jk}. \tag{39}$$

The Nash solution can again be obtained by using reaction curves: see Fig. 2(b).

The effect of allocated quotas is obvious: producer 2's capacity level $E_2^{\max}$ has very little influence on producer 1's optimal capacity. For N equivalent producers we obtain the equation

$$\phi(N-1)\bar{E}) = \bar{E}$$

for the Nash solution $E_i^{\max} = \bar{E}$, where ϕ is the individual producer's reaction function. If $\phi(E) \sim E_0$ for large E [Fig. 2(b)], this yields $\bar{E} \sim E_0$ for large N. Thus by appropriately fixing E_0 (by means of Q) and N, the manager can ensure the generation of significant economic rents. The distribution of these rents will depend on the conditions under which quotas are distributed to producers.

The reader is referred to Christy (Ref. 21) for further discussion of the advantages and disadvantages of allocated quota systems. Let me note one possible objection: fishermen possessing a fixed quota may be prevented from benefiting from a lucky catch. However, it seems that in practice, quota schemes could be devised with sufficient flexibility to accommodate such possibilities.

6. Conclusion

It has been the purpose of this article to introduce a rigorous analytical framework for the discussion of problems pertaining to the economic rationalization of common-property fisheries. While many important aspects of the problem (such as alternative objectives, stochastic effects, uncertainty, distributional and political problems) have been ignored, the results of the analysis seem to be quite robust. The incentives leading to the dissipation of economic rent under competitive conditions of exploitation cannot be overcome by the traditionally accepted methods of regulation. Only two approaches, royalties on catch (or effort) and allocated quotas, seem to offer any hope of effectively thwarting this incentive.

The discussion has not allowed for the possibility of cooperation between producers. With the possible exception of a few fisheries exploited exclusively by the "factory" fleets of the USSR and other Eastern bloc countries, individual producers will normally be too numerous to allow a cooperative scheme to operate stably in practice. The theory of cooperative games may prove useful, however, in analyzing the bargaining possibilities at the international level (Ref. 18).

To conclude, I will express the hope that further application of the mathematical theory of games will lead to a deeper understanding of these interesting and important problems.

Acknowledgment

Research partially supported by the National Research Council of Canada, Grant A3990.

References

1. Hardin, G., The tragedy of the commons, *Science*, Vol. 162, pp. 1243–1247, 1968.
2. Christy, F. T., Jr and Scott, A. D., *The Common Wealth in Ocean Fisheries*, Johns Hopkins University Press, Baltimore, Md., 1965.
3. Clark, C. W., *Mathematical Bioeconomics: The Optimal Management of Renewable Resources*, Wiley-Interscience, New York, 1976.
4. Scott, A. D. and Southey, C., The problem of achieveng efficient regulation of a fishery, *Economics of Fisheries Management: A Symposium*. Edited by A. D. Scott, University of British Columbia, Institute of Animal Research and Ecology, Vancouver, B.C., 1970.

5. GULLAND, J. A., *The management of Marine Fisheries*, University of Washington Press, Seattle, Wash., 1974.
6. GORDON, H. S., Economic theory of a common-property resource: The Fishery, *Journal of Political Economy*, Vol. 62, pp. 124–142, 1954.
7. ANDERSON, L. G., *The Economics of Fisheries Management*, Johns Hopkins University Press, Baltimore, Md., 1977.
8. CLARK, C. W. and MUNRO, G. R., The economics of fishing and modern capital theory: A simplified approach, *Journal of Environmental Economics and Management*, Vol. 2, pp. 92–106, 1975.
9. PLOURDE, C. G., Exploitation of common-property replenishable resources, *Western Economics Journal*, Vol. 9, pp. 256–266, 1971.
10. QUIRK, J. P. and SMITH, V. L., Dynamic economic models of fishing, *Economics of Fisheries Management, A Symposium*. Edited by A. D. Scott, University of British Columbia, Institute of Animal Research and Ecology, Vancouver, British Columbia, 1970.
11. SMITH, V. L., On models of commercial fishing, *Journal of Political Economy*, Vol. 77, pp. 181–198, 1969.
12. ISAACS, R., Differential games, their scope, nature, and future, *Journal of Optimization Theory and Applications*, Vol. 3, pp. 283–295, 1969.
13. SCHAEFER, M. B., A study of the dynamics of the fishery for yellowfin tuna in the eastern tropical Pacific Ocean, *Bulletin of the Inter-American Tropical Tuna Commission*, Vol. 2, pp. 247–285, 1957.
14. SPENCE, M. and STARRETT, D., Most rapid approach paths in accumulation problems, *International Economic Review*, Vol. 16, pp. 388–403, 1975.
15. CLARK, C. W., CLARKE, F. H., and MUNRO, G. R., The optimal exploitation of renewable resource stocks: Problems of irreversible investment, *Econometrica*, Vol. 47, pp. 25–47, 1979.
16. HO, Y. C., Differential games, dynamic optimization, and generalized control theory, *Journal of Optimization Theory and Applications*, Vol. 6, pp. 179–209, 1970.
17. NASH, J. F., Non-cooperative games. *Annals of Mathematics*, Vol. 54, pp. 286–295, 1951.
18. MUNRO, G. R., The optimal management of transboundary renewable resources, *Canadian Journal of Economics*, Vol. 12, pp. 355–376, 1979.
19. CASE, J., A dynamic version of Cournot's problem, *Differential Games and Control Theory*. Edited by E. O. Roxin, P. T. Liu, and R. L. Sternberg, Marcel Dekker, New York, pp. 161–178, 1977.
20. BROWN, G., Jr., An optimal program for managing common property resources with congestion externalities, *Journal of Political Economy*, Vol. 82, pp. 163–174, 1974.
21. ADASIAK, A., Alaska's experience with limited entry, *Journal of the Fisheries Research Board of Canada*, Vol. 36, pp. 770–782, 1979.
22. CHRISTY, F. T., Jr., Limited access systems under the Fishery Conservation and Management Act of 1976, *Economic Impacts of Extended Fisheries Jurisdiction*. Edited by L. G. Anderson, Ann Arbor Science, Ann Arbor, Mich., 1976.

8

Cartel and Oligopoly Pricing of Nonreplenishable Natural Resources

TRACY R. LEWIS AND RICHARD SCHMALENSEE

1. Introduction

Few real-world market structures correspond well to either of the textbook polar cases of absolute monopoly and perfect competition. Formal models of two intermediate structures have been thoroughly and usefully analyzed under static conditions.† The first model assumes that a single firm or a stable cartel controls a sizable fraction of industry capacity and faces a competitive fringe of many small suppliers, each too tiny to have any noticeable effect on market price. Fringe members thus take price as beyond their control and choose output to maximize profit. The dominant firm or cartel maximizes its profit subject to the constraint imposed by fringe supply behavior. The second model deals with non-cooperative or Cournot–Nash oligopoly. A finite number of sellers is assumed, each large enough to have some control over price. Market equilibrium is defined as a situation in which no individual seller can increase its profits by changing only its own output, given the outputs of the other sellers.

† See, for instance, Scherer (Ref. 1, Chap. 5). It is fair to say that these are the only widely used formal models of intermediate structures except for those concerned with entry, cartel stability, or differentiated products.

TRACY R. LEWIS • Department of Economics, University of Arizona, Tucson, Arizona. RICHARD SCHMALENSEE • Sloan School of Management, Massachusetts Institute of Technology, Cambridge, Massachusetts.

This essay is concerned with the implications of these structures in markets for nonrenewable natural resources.† Following Hotelling (Ref. 5) and numerous subsequent authors, we assume that the total reserves of the resource in the hands of each producer cannot be increased and are reduced by production. Demand and cost conditions, including the relevant rate of interest, are constant over time. In such a world, producers must rationally consider price or output paths over time, so that both textbook models become nonzero-sum differential games. In what follows, we examine solutions to the games implied by various assumptions.

Our work relies heavily on that of Salant (Ref. 6), who seems to have done the only theoretical analysis of either incomplete cartel or noncooperative oligopoly in nonreplenishable resource markets.‡ Section 2 presents and extends his model of cartel equilibrium. Following Salant, we assume that the cartel and the competitive fringe have and use perfect information about prices, outputs, and reserves of all producers at each instant. Under this assumption, it is clear that all expectations must be realized.

The assumption of perfect information seems a bit unrealistic in this context, however. At the very least, it is difficult for suppliers to verify rivals' assertions about levels of reserves. Accordingly, Section 3 considers the implications of relaxing the assumption that the competitive fringe knows the cartel's reserves. The cartel can lie about reserves, thus creating expectations about future prices and outputs that will later prove to have been incorrect. (The fringe suppliers are assumed to tell the truth, perhaps because they are too small to lie profitably.) However, we suppose that the fringe can monitor cartel sales. Thus, in order for a lie about reserves to be believed, the cartel must follow the output path that would be optimal for the announced reserve level. We derive the surprising result that the cartel's optimal policy is to tell the truth. Competitive monitoring of output (plus careful analysis of observed output decisions) is as effective a constraint on cartel behavior as full information about reserves.

Section 4 presents some exploratory analysis of noncooperative oligopoly in natural resource markets, assuming perfect information. We find that if all sellers have equal reserves and equal extraction costs, then

† Interesting analyses of oligopoly pricing of renewable resources appear in Clark (Ref. 2) and Levhari and Mirman (Ref. 3). Stiglitz (Ref. 4) has analyzed the relation between competitive and monopoly behavior in nonrenewable resource markets.

‡ In two interesting and important recent papers, Pindyck (Refs. 7, 8) computes the wealth-maximizing price policies for a number of actual incomplete cartels, taking into account distributed lags in demand. To obtain tractable models, however, Pindyck assumes that competitors of these cartels behave myopically. In both Salant's (Ref. 6) analysis and the models that follow, all producers are fully rational, but demand curves are stationary.

existence and uniqueness of equilibrium can be established and some comparative dynamic results obtained.

As will become clear, some of our results depend on assumptions that are stronger than we would like. Moreover, many potentially important aspects of natural resource markets are missing entirely from our models: strategic or other changes in buyer behavior, exploration for new reserves, differences in cartel members' objectives, cartel stability problems, and others.† Much work remains to be done on imperfect competition in nonrenewable natural resource markets.

2. The Basic Cartel Model with Full Information

2.1. Assumptions

Following Salant (Ref. 6), we consider a market for a nonreplenishable natural resource (such as oil or bauxite) in which consumers purchase supplies at the same price from independent firms. Market transactions are described in terms of a Cournot–Nash, noncooperative game. Each supplier maximizes its own discounted profits, taking as given the sales path for all other firms. One group of firms that can presumably exercise some market power by virtue of its large and concentrated resource holdings forms a cartel that acts collusively to set prices or limit sales. The remainder of the world's resource stock is dispersed among a sufficiently large number of other suppliers so that a dominant firm model results. Small producers act like competitors who perceive prices as given in each time period and choose a sales path to maximize discounted profits. The cartel simultaneously chooses a price path, supported by cartel sales, to maximize discounted profits taking aggregate sales by the competitive fringe as given. An equilibrium sales or price path is said to exist whenever the cartel and each fringe supplier are simultaneously maximizing individual discounted profits while taking as given the sales path of all other firms.

Let $q(t)$ be total market sales and $q^m(t)$ and $q^c(t)$ be the sales by the cartel and competitive fringe, respectively, at time t. The stationary inverse demand function for the resource $p(q)$, with $p'(q) < 0$, is characterized by whether elasticity $\varepsilon(q)$ is decreasing, constant, or increasing as a function of q:

$$\text{(DE)}\quad \varepsilon'(q) < 0;\ \varepsilon(\bar{q}) = 1;\ p(0) < \infty.$$
$$\text{(CE)}\quad \varepsilon'(q) = 0;\ \varepsilon(q) > 1;\ (\forall q).$$
$$\text{(IE)}\quad \varepsilon'(q) > 0;\ \varepsilon(q) > 1;\ (\forall q).$$

† For work on these and related aspects of resource markets, see Eckbo (Ref. 9), Hnyilicza and Pindyck (Ref. 10), Kosbud and Stokes (Ref. 11), Pakravan (Ref. 12), Pindyck (Refs. 7, 8), and Schmalensee (Ref. 13).

Salant (Ref. 6) employs the (DE) specification which assumes a finite "choke price" $p(0)$ and decreasing demand elasticity. The alternatives (CE) and (IE) are included here for comparison with (DE).† Specifications (DE) and (CE) are fairly standard; (IE) corresponds to a demand schedule with elasticity increasing in consumption. A justification for this assumption is that for small quantities, demand may be inelastic if certain amounts of the resource are essential in the production of some goods. At lower prices, however, the resource may be used in other industries for which substitute inputs exist as well. Consequently, the elasticity of aggregate demand may increase.

Let I, I^m, and I^c represent initial total market, cartel, and competitive fringe reserves; then the basic model employs a full information assumption.

(I1) All suppliers can observe $\{q(t), q^m(t), q^c(t), I^m, I^c\}$. Assumption I1 is modified in Section 3 to allow for partial observability of I^m by fringe extractors.

2.2. Characterization of Equilibria

Market equilibrium occurs when individual suppliers are each maximizing discounted profits, taking the sales paths of all other extractors as given. This Nash equilibrium concept is particularly compelling if one assumes the existence of a futures market for the resource, where extractors can contract for a stream of future shipments of other firms. For simplicity, zero marginal extraction costs are assumed. (Results presented here are basically unchanged with the introduction of constant positive marginal extraction costs. Details are omitted.) Taking the sales paths of other firms as given, straightforward maximization of cartel and individual fringe supplier profits yield the necessary conditions (1)–(4) for an equilibrium.‡

$$e^{-rt}\mathrm{MR}^m(q^m, q^c) \equiv e^{-rt}\left[p(q^m + q^c) + \frac{dp}{dq}q^m\right] = \mathrm{MR}_0^m, \tag{1}$$

$$\int_0^T q^m(t)\,dt = I^m, \tag{2}$$

$$e^{-rt}p(q^m + q^c) = p_0 \quad \text{whenever } q^c > 0, \tag{3}$$

$$\int_0^S q^c(t)\,dt = I^c. \tag{4}$$

† For specifications (CE) and (IE) choke prices are infinite so that we need to assume $\lim_{q\to 0} p(q)q < \infty$.

‡ These are equivalent to (1)–(4) in Salant (Ref. 6). Henceforth, time arguments are deleted where no confusion exists.

Given the competitive sales path, the cartel maximizes profits by equating discounted marginal revenue (from its excess demand curve) across all time periods as in equation (1). Equation (3) is an equilibrium arbitrage condition which must hold whenever price-taking competitive fringe suppliers operate in the market. Inventory constraints are given by (2) and (4), where T and S represent the extraction horizon for the cartel and competitive fringe, respectively.

The first thing to notice about this equilibrium is that $q^m(t) > 0$ whenever $q(t) > 0$. This stems from the fact that $\mathrm{MR}^m(0, q^c) = p(q)$ so that q^m can never go from a positive to a zero value or from a zero to a positive value while $q(t) > 0$, without violating (1) and (3).

We need two more preliminary results. Let $\{\tilde{q}^m(t)\}$ and $\{\tilde{q}^c(t)\}$ denote the paths that maximize discounted profits for the cartel and that would result in competitive equilibrium, respectively, in the extreme cases where $I^m = I$ and $I^c = I$. The paths for the cartel and competitive fringe are then described by conditions (1)–(2) and (3)–(4), respectively (making the obvious adjustments that $q^c(t) = 0\,(\forall t)$ if $I^m = I$, and so forth, where necessary). If, in addition to (DE), (CE), and (IE), we assume that $\mathrm{MR} = R'(q) > 0$, $R''(q) < 0$, where $R(q) = p(q)q$, we can establish the following lemma.

Lemma 2.1. Considering $\dot{\tilde{q}}^c(q)$ and $\dot{\tilde{q}}^m(q)$ as functions of q, we have $\dot{\tilde{q}}^c, \dot{\tilde{q}}^m < 0$, and

$$\dot{\tilde{q}}^c(q) \lesseqgtr q^m(q) \qquad \text{as } \varepsilon'(q) \lesseqgtr 0, q > 0, \tag{5}$$

$$\frac{\dot{\mathrm{MR}}}{\mathrm{MR}}(q) \lesseqgtr \frac{\dot{p}}{p}(q) \qquad \text{as } \varepsilon'(q) \lesseqgtr 0, q = q^m(t). \tag{6}$$

Lemma 2.2

$$\tilde{q}^m(0) \gtreqless \tilde{q}^c(0) \qquad \text{as } \varepsilon'(q) \gtreqless 0, \tag{7}$$

where $\mathrm{MR} = p + (dp/dp)q$.

Proofs of these lemmas follow readily by taking time derivatives of (1) and (3).

Using Lemmas 2.1 and 2.2, we see that the Nash equilibrium sales paths for various demand specifications are as follows. Let $s(t) = I^m(t)/I(t)$ represent current cartel inventories as a fraction of total world inventories.

Case (DE). For $s(t) = 1$ or 0, we revert to the polar cases just discussed. Suppose that $0 < s(0) < 1$. Whenever $q^c(t) = 0$, then $\mathrm{MR}^m = \mathrm{MR}$ and (1) and (6) imply that $\dot{p}/p < r$. Consequently, fringe suppliers never delay the beginning of extraction since real prices are falling. From (1) and (3), $\dot{\mathrm{MR}}^m/\mathrm{MR}^m = \dot{p}/p$, or $\mathrm{MR}^m = Kp$, $0 < K < 1$, whenever $q^m, q^c > 0$. Solving for q^c yields

$$q^c = q[1 - (1 - K)\varepsilon(q)]. \tag{8}$$

The existence of a finite choke price implies $\varepsilon(q) \to \infty$ as $q \to 0$. Consequently, by (8), q^c [and $I^c(t)$] go to zero while q, q^m, and I^m are still positive. Thus, initially both the competitive fringe and the cartel operate for a period during which prices rise at the rate of interest. Eventually, fringe supplies are exhausted, and the entire market is left to the cartel, which then operates as a monopoly with $\dot{p}/p < r$.

Case (CE). With $\varepsilon' = 0$, it is easy to verify from (8), (2), and (4) that $q^c(t) = [I^c(0)/I(0)]q$, $q^m(t) = [I^m(0)/I(0)]q$, and $\dot{\mathrm{MR}}^m/\mathrm{MR}^m = \dot{p}/p = r$ for all t. For given initial world reserves of the resource $I(0)$, the equilibrium price and total sales path are independent of the distribution of resources between the cartel and fringe suppliers. The condition $\varepsilon' = 0$ implies $p(0) = \infty$, so that resources are allocated over infinite time at a constant real price.

Case (IE). Whenever $q^c(t) = 0$, (1) and (6) imply $\dot{p}/p > r$. Fringe suppliers delay extraction as much as possible, consistent with the eventual sale of all their inventories. Of course, real (discounted) prices can rise only if speculators cannot buy resources and store them for later sale. This is probably a fair assumption in the case of oil. Once fringe extraction begins, (8) and the condition $\varepsilon'(q) > 0$ imply that $q^c(t)$ is positive for $q > 0$. Since $\varepsilon'(q) > 0$, we have $p(0) = \infty$ so that resources are allocated over infinite time. In summary, if I^m is sufficiently large relative to I^c, there may exist an initial period during which ($q^c = 0$, $q^m > 0$) and real prices rise.† Eventually, the fringe suppliers enter the industry, whereupon prices rise at a competitive rate. Summarizing, in the instance of (DE), the market is initially characterized by competitive price movements, followed by a second period of monopoly control. The opposite sequence of events can occur in case (IE). For cases (DE) and (IE), market equilibriums become more monopolistic in character as $s = I^m/I \to 1$, whereas the equilibrium price and total sales path are independent of s for the (CE) specification.

† Clearly, for the extreme case where $I^m = I$, the entire stock is allocated in a monopoly market and $\dot{p}/p > r$. If we change the distribution of initial reserves by decreasing I^m and increasing I^c, we eventually encounter a situation in which all reserves are allocated under competitive conditions with $\dot{p}/p = r$. Recall from Lemma 2.2 that $\tilde{q}^c(0)$ is the initial sales rate that occurs when all resources are sold competitively with $\dot{p}/p = r$. When ($q^c, q^m > 0$) we have $q^m = (1 - K)\varepsilon(q)q$ by (8). Thus the largest value of I^m consistent with the competitive sale of all reserves is $\bar{I}^m = \int_0^\infty (1 - K)\varepsilon(q)q\,dq$, where $q(0) = \tilde{q}^c(0)$ and $K = 1 - 1/\varepsilon(\tilde{q}^c(0))$. For $I^m > \bar{I}^m$, an initial period exists in which $q^c = 0$, $q^m > 0$.

2.3. Existence and Uniqueness of Equilibrium

This section is devoted to establishing the following basic result, which generalizes the existence proof in Salant (Ref. 6).

Proposition 2.1. Given (DE), (CE), (IE), and (I1), a unique equilibrium exists characterized by (1)–(4).

Proof. Proof for the case (DE) is presented here, noting that the other cases are established by a similar argument.

a. The inventory constraint for total reserves is written

$$\hat{I}(0, S) \equiv \int_0^S q(t)\,dt + \int_S^T q(t)\,dt = I. \tag{9}$$

During the first interval, $(0, S)$ resources are allocated competitively with q^m, $q^c > 0$, and $\dot{p}/p = r$; during the second interval, the allocation is monopolistic with $q^c > 0$ and $\dot{p}/p < r$. Changing the variable of integration in (9), we obtain

$$\hat{I}(q_0, q_1) = \int_{q_1}^{q_0} (-q/\dot{\tilde{q}}^c)\,dq + \int_0^{q_1} (q/\dot{\tilde{q}}^m)\,dq = I(0), \tag{10}$$

where $\dot{\tilde{q}}^c$ and $\dot{\tilde{q}}^m$, calculated from (3) and (1), respectively, represent the competitive and cartel-dominated rates of change in sales as a function of q, and $q(0) = q_0$, $q(S) = q_1$, and $q(T) = 0$.

b. Rewriting the stock constraints for the case in which all resources are either sold monopolistically or competitively, we obtain

$$\int_0^{\tilde{q}^m(0)} (-q/\dot{\tilde{q}}^m)\,dq = I, \tag{11}$$

$$\int_0^{\tilde{q}^c(0)} (-q/\dot{\tilde{q}}^c)\,dq = I. \tag{12}$$

Then from Lemmas 2.1 and 2.2 and from (11) and (12),

$$\hat{I}(q_0, q_1) > \int_{q_1}^{q_0} (-q/\dot{\tilde{q}}^c)\,dq + \int_0^{q_1} (-q/\dot{\tilde{q}}^c)\,dq \geq I(0) \tag{13}$$

whenever $q_0 \geq \tilde{q}^c(0)$, and

$$\hat{I}(q_0, q_1) > \int_{q_1}^{q_0} (-q/\dot{\tilde{q}}^m)\,dq + \int_0^{q_1} (-q/\dot{\tilde{q}}^m)\,dq \leq I(0) \tag{14}$$

whenever $q_0 \leq \tilde{q}^m(0)$, so that in order to satisfy (10)–(12), we must have

$$\tilde{q}^m(0) < q_0 < \tilde{q}^c(0). \tag{15}$$

c. From (10) we can represent q_1 as a function of q_0. Note that Lemmas 2.1 and 2.2 and (11) and (12) imply

$$\hat{I}(q_0, q_0) > I > \hat{I}(q_0, 0). \tag{16}$$

Differentiating equation (10) with respect to q_1 yields

$$\frac{d\hat{I}}{dq_1} = q/\dot{\tilde{q}}^c - q/\dot{\tilde{q}}^m > 0. \tag{17}$$

Together, (16) and (17) imply that for each q_0 there is a unique $q_1(q_0)$ determined by (10).

d. From c we can show that there is a one-to-one relationship between I^c and q_0 for a given $I(0)$. Equations (1) and (3) imply that $q^c(q_1) = 0$. From (8) we have

$$q^c = [1 - (1 - K)\varepsilon(q)]q$$

which, together with the $q^c(q_1) = 0$ condition, implies that

$$K = 1 - 1/\varepsilon(q_1(q_0)).$$

We can express I^c in terms of q_0 as

$$I^c(q_0) = \int_{q_1(q_0)}^{q_0} -[1 - (1 - K)\varepsilon(q)]q/\dot{\tilde{q}}^c \, dq. \tag{18}$$

Note that $q_1(\tilde{q}^c(0)) = 0$ and $q_1(\tilde{q}^m(0)) = \tilde{q}^m(0)$ by (10)–(12). Then from (18), assuming that $s \in (0, 1)$, we obtain

$$I^c(\tilde{q}^c(0)) = I(0) > 0 = I^c(\tilde{q}^m(0)). \tag{19}$$

and differentiating (18) with respect to q_0 yields

$$\frac{dI_c}{dq_0} = \frac{q[1 - (1 - K)\varepsilon(q_0)]}{\dot{q}(q_0)} + \int_{q_1}^{q_0}\left(-\frac{dK}{dq_1}\frac{dq_1}{dq_0}\Big/\dot{q}\right)\varepsilon(q)\,dq > 0. \tag{20}$$

Together (19) and (20) imply that there exists a unique q_0 for each I^c and, conversely, that for each q_0 there is a unique I^c.

In summary, given initial stocks (I, I^c), there exists a unique q_0 satisfying (18). But q_0 maps into a unique value for q_1 to satisfy (10). Given q_0 and q_1, the equilibrium is completely described by (1), (3), (11), and (12), and the proof of Proposition 2.1 is complete. □

2.4. Properties of Equilibrium

Let p_0 and MR_0^m denote the constant discounted price and marginal revenue accruing to fringe suppliers and the cartel, respectively, in market

Table 1. Signs of Comparative Static Derivatives

Endogenous variables	Exogenous variables		$s = I^m/I$ [a]		
	I^c	I^m	A	B	C
MR_0^m	−	−	−	(0)	−
p_0	−	−	+	(0)	−
W^m	−	+	+	(0)	+
W^c	+	−	−	(0)	−
$W^m + W^c$	+	+	+	(0)	+
W^m/I^m	−	−	+	(0)	+
W^c/I^c	−	−	+	(0)	+

[a] Column A shows the comparative static derivatives for the (DE) case. In column B, the zeros in parentheses indicate that comparative static derivatives are zero in those cases (CE) and sometimes (IE) where changes in s do not affect the price and total quantity paths in equilibrium. Column C shows the comparative static derivatives for the (IE) case where changes in s affect price and total quantity paths in equilibria. Results are derived assuming $(d/dq)\mathrm{MR}(q) < 0$.

equilibrium, and let $W^m(I^m, I^c)$ and $W^c(I^m, I^c)$ represent total discounted profits earned by the cartel and competitive fringe sector. Then the qualitative implications of the market equilibrium (1)–(4) for the comparative static effects of changes in initial inventory levels are presented in Table 1. Details of the sign derivations for the first five rows of Table 1, derived following the methods of Hartwick (Ref. 14), are presented in Appendix A for later use.

Lines 1 and 2 of Table 1 indicate that cartel and competitive fringe marginal value of reserves, measured by MR_0^m and p_0, respectively, decline with greater initial inventories. Total profits for each sector vary directly (inversely) with own initial reserves (reserves of the other sector), as revealed in lines 3 and 4. The last three lines of Table 1 imply that total industry profits and individual profits for all suppliers increase as S increases whenever equilibrium prices are affected by the division of initial reserves between the two sectors. The larger the cartel is relative to the market, the better off are all suppliers in aggregate.

Thus, the formation of a cartel benefits all extractors, including the fringe suppliers. Yet one can show that fringe suppliers can generally earn greater profits than cartel members, that individual members can profitably defect from the cartel, assuming others maintain the cartel discipline, and that cartels are therefore basically unstable here, as they are in static textbook models. The fringe supplier is able to sell his output at some constant real price p_0. Differentiating (1) and (3) with respect to time yields

$$\dot{p}/p \gtreqless r \qquad \text{as } \varepsilon' \gtreqless 0 \text{ for } q^c = 0, q^m > 0,$$

$$\dot{p}/p = r \qquad \text{for } q^c, q^m > 0,$$

implying that for cases DE (IE), the cartel must sell some of its stock during a final (initial) extraction phase at discounted values below p_0. Thus, the average present value of cartel reserves is less than that of reserves held by competitors. In the constant elasticity case, individual extractors receive the same discounted price, independent of the distribution of initial reserves.

3. A Cartel Model with Partial Observability of Initial Inventories

3.1. Introduction

In the previous section, market strategies were examined for a game with complete information, in which each of the suppliers knew the sales and initial inventories of all the others. Because there is significant uncertainty about the world supply of some resources like oil, it seems appropriate to consider situations in which there is partial knowledge of initial inventories. In this section, we assume that (a) initial competitive fringe supplies are known by all extractors, but cartel inventories are observed only by the cartel, and (b) for strategic purposes, the cartel can misrepresent its initial inventories to fringe suppliers so long as the cartel's sales path is "consistent" with its announced inventory levels. The case considered here gives rise to a sophisticated–naive, leader–follower model, reminiscent of the static Stackelberg model. The cartel chooses a best inventory and sales strategy taking the response of the fringe suppliers into account. The justification for treating the two sectors differently is that individual fringe suppliers perceive that they cannot alter prices by acting strategically, whereas the cartel recognizes its ability to affect prices by announcing different inventory and sales paths. The model of cartel pricing under incomplete information analyzed in this section is characterized by the following assumption.

(I2) Cartel suppliers can observe $q^c(t)$, I^c.

Competitive fringe suppliers can observe $q^m(t)$ and the following market procedure:

(P1) At time t_0, the cartel announces a level of initial stock $\bar{I}^m$.†

† Actually, it is sufficient for the cartel to just announce a quantity path, since its total initial reserves can be inferred from the sales path.

Under the assumption that I^c is known, trade for current and future sales of the resource by both continues until the sales paths $q^m(\bar{I}^m, I^c, \tau)$ and $q^c(\bar{I}^m, I^c, t)$ satisfying the conditions (1)–(4) for a market equilibrium are announced by the cartel and competitive fringe sectors. From Proposition 2.1 we know that these paths exist and are unique for each pair $(\bar{I}^m, I^c)$. The market remains in equilibrium, and fringe extractors continue to sell equilibrium quantities $q^c(\bar{I}^m, I^c, t)$ based on the assumption that $\bar{I}^m$ is the true cartel inventory level, until the cartel deviates from its equilibrium path. At this point, the cartel is forced to reveal its true current inventory level, and the market comes to a new equilibrium based on actual reserve levels.

Given (P1), one cartel strategy is to announce an initial stock $\bar{I}^m$ which is larger than the actual cartel inventory. Once a market equilibrium is established based on initial stocks $(\bar{I}^m, I^c)$, resources will initially be consumed at a faster rate $(\partial q_0/\partial I^m > 0)$ since stocks are believed to be large. If, as a result, fringe suppliers are induced to sell most of their stocks in early time periods, the cartel can assume a monopoly position later, when fringe supplies have been depleted. Note that the cartel is allowed to change sales without compensating consumers who might be harmed in the process. This would not be possible if future contracts for sales were perfectly binding. Of course, in order to be convincing, the cartel must sell certain quantities $q^m(\bar{I}^m, I^c, t)$ at artificially deflated prices in return for the monopoly position that it hopes to attain in the future. Obviously, this is just one of several strategies that the cartel might pursue.

3.2. Cheating by Misrepresenting Initial Inventories

This procedure (P1) describes the simplest option for reporting initial inventories. Clearly, the strategy in which the cartel reports its actual initial inventory and then follows the market equilibrium path thereafter is feasible under (P1), and the following proposition asserts that this is in fact the optimal policy for the cartel to choose, even for more complicated strategies. For simplicity in the remainder of the paper, we shall confine our attention to the decreasing elasticity case. To make some of the mathematics more tractable, we consider a special case of (DE):

$$\text{(DE}'\text{)}\quad p(q) = a - q^{\alpha+1}; \qquad -1 < \alpha < \infty,\ a > 0.$$

The (DE′) demand specification is reasonably general, for it allows for convex and concave demand schedules.

Proposition 3.1. Given the partial information structure (I2) and the demand specification (DE′), the best strategy for the cartel is to announce its actual initial inventories and follow the equilibrium path.

Proposition 3.1 is quite general. It rules out the optimality of any strategy that involves a single or repeated misrepresentation of existing cartel reserves.

Proof. First we demonstrate that any cheating strategy the cartel pursues that satisfies procedure (P1) is dominated by telling the truth. All cheating strategies must involve a period of time in which a strategy of the type described by (P1) is used, and we are therefore able to rule out the optimality of all possible cheating schemes.

a. At some time t_0 the cartel announces that its current inventories are $I^m(t_0)$. We arbitrarily set $t_0 = 0$. Assume that I^m is the actual initial cartel stock. Construct

$$V(\bar{I}^m, I^c, \tau) = \int_0^\tau e^{-rt}[p_1(q_1^m, q_1^c, t)q^m]\,dt + e^{-rt}W^m\left(I^m - \int_0^\tau q_1^m(\bar{I}^m, I^c, t)\,dt, I^c - \int_0^\tau q_1^c(\bar{I}^m, I^c, t)\,dt\right), \quad (21)$$

where q_1^m and q_1^c are the unique equilibrium cartel and competitive fringe sales and p_1 is the equilibrium price, all as functions of $\bar{I}^m$, I^c, and t. The function $V(\cdot)$, is the discounted profit to the cartel which announces initial inventories $\bar{I}^m$, follows the Nash equilibrium corresponding to $(\bar{I}^m, I^c)$ from time 0 to τ, then announces its actual inventories at time τ, which are $I^m - \int_0^\tau q_1^m(t)\,dt$, and receives the payoff $W(\cdot, \cdot)$, which is the sum of discounted profits earned by the cartel for a Nash equilibrium corresponding to existing inventories $(I^m(T), I^c(T))$.

b. Differentiating $V(\bar{I}^m, I^c, T)$ with respect to τ, we obtain

$$\frac{\partial V(\bar{I}^m, I^c, \tau)}{\partial \tau} = \left[P_1(\tau)q_1^m(\tau) - rW^m(I^m(\tau), I^c(\tau)) - \frac{\partial W^m}{\partial I^m}q_1^m(\tau) - \frac{\partial W^m}{\partial I^c}q_1^c(\tau)\right]e^{-r\tau}. \quad (22)$$

We wish to show $\partial V/\partial \tau < 0$ for $\bar{I}^m(0) \neq I^m(0)$.

For $\bar{I}^m(0) = I^m(0)$ we have the identity

$$V(I^m, I^c, \tau) = W^m(I^m(0), I^c(0)) \qquad (\forall T) \quad (23)$$

so that

$$\frac{\partial V(I^m, I^c, \tau)}{\partial \tau} = 0 \qquad (\forall \tau). \quad (24)$$

In Appendix A we show that $\partial W^m(\tau)/\partial I^m(\tau) = \mathrm{MR}_2^m(\tau)$ and $\partial W^m(\tau)/\partial I^c(\tau) = \mathrm{MR}_2^m(\tau) - P_2(\tau)$ where $P_2(\tau)$ and $\mathrm{MR}_2^m(\tau)$ are the constant discounted price and cartel marginal revenue for the new Nash equilibrium corresponding to initial inventories $(I^m(\tau), I^c(\tau))$. Substituting for these terms in (24) yields

$$rW(I^m(\tau), I^c(\tau)) = P_2(\tau) - \mathrm{MR}_2^m(\tau)q_2(\tau). \tag{25}$$

Substituting for $rW(I^m(\tau), I^c(\tau))$ in (22) yields

$$\begin{aligned}\frac{\partial V(I^m, I^c, \tau)}{\partial \tau} &= \{P_1(\tau)q_1^m(\tau) - [P_2(\tau) - \mathrm{MR}_2^m(\tau)]q_2(\tau) \\ &\quad - \mathrm{MR}_2^m(\tau)q_1^m(\tau) - [\mathrm{MR}_2^m(\tau) - P_2(\tau)][q_2(\tau) - q_2^m(\tau)]\}e^{-r\tau} \\ &= \{[P_1(\tau) - P_2(\tau)]q_1^m(\tau) + [P_2(\tau) - \mathrm{MR}_2^m(\tau)][q_1(\tau) - q_2(\tau)]\}e^{-r\tau}. \end{aligned} \tag{26}$$

From (26) we can verify that $\partial V(I^m, I^c, \tau)/\partial\tau = 0$ by noting that $P_1(\tau) = P_2(\tau)$ and $q_1(\tau) = q_2(\tau)$ (equilibrium price and quantity paths do not change) when $\bar{I}^m(0) = I^m(0)$.

c. We now wish to show that

$$\frac{\partial V(\bar{I}^m, I^c, \tau)}{\partial \tau} < 0 \qquad \text{for } \bar{I}^m \neq I^m. \tag{27}$$

Suppose that the demand schedule is linear with $\alpha = 0$ and $dp/dq = -1$. The second expression on the right-hand side of (26) can be rewritten as

$$\begin{aligned}[P_2(\tau) - \mathrm{MR}_2^m(\tau)][q_1(\tau) - q_2(\tau)] &= \left(-\frac{dp}{dq}q_2^m(\tau)\right) \\ &\quad \times [q_1(P_1(\tau)) - q_2(P_2(\tau))] = -[P_1(\tau) - P_2(\tau)](q_2^m(\tau)), \end{aligned} \tag{28}$$

which, when combined with (26), yields

$$\frac{\partial V}{\partial \tau} = [P_1(\tau) - P_2(\tau)][q_1^m(\tau) - q_2^m(\tau)], \qquad \alpha = 0. \tag{29}$$

Similarly, one can establish the following inequalities in the case (DE′) when $\alpha \neq 0$:

$$\frac{\partial V}{\partial \tau} < [P_1(\tau) - P_2(\tau)][q_1^m(\tau) - q_2^m(\tau)], \qquad \alpha \in (0, \infty), \tag{30}$$

$$\begin{aligned}\frac{\partial V}{\partial \tau} &< [P_2(\tau) - \mathrm{MR}_2^m(\tau)][q_1(P_1(\tau)) - q_2(P_2(\tau))] \\ &\quad \times \left(\frac{q_2^m(\tau) - q_1^m(\tau)}{q_2^m(\tau)}\right), \qquad \alpha \in (-1, 0). \end{aligned} \tag{31}$$

In Appendix B we establish

$$q_1^m(\tau) \gtreqless q_2^m(\tau) \qquad \text{as } \bar{I}^m \gtreqless I^m(0), \tag{32}$$

which, together with the condition $\partial q(I^m, I^c, 0)/\partial I^m > 0$ is sufficient to establish

$$\frac{\partial V}{\partial \tau} < 0 \qquad \text{whenever } \bar{I}^m \neq I^m. \tag{33}$$

d. Thus, telling the truth dominates any strategy involving an initial period of cheating by stock misrepresentation. This result generalizes to more complicated strategies. Suppose that the cartel announces different inventories at N various times, assuming fringe suppliers naively continue to adjust sales in accord with these inventories. From (33) working backward, the best policy for the cartel is to reveal its true inventory in the last period N as well as in period $N - 1$. But the same reasoning that led to (33) implies that the cartel should reveal its true stock level in period $N - 2$, given that actual inventories are to be announced for the periods $N - 1$ and N. Repeating the argument for periods $N - 3, N - 4, \ldots,$ and so on, we can show that telling the truth dominates any cheating strategy, and Proposition 3.1 is established. □

When cartel sales can be monitored, cheating does not work because the cartel must invest too much to convince fringe suppliers that its announced inventories are valid. For example announcing very large inventories $\bar{I}^m$ in the hopes of increasing initial sales and inducing fringe suppliers to sell out early requires that the cartel sell substantial quantities at artificially deflated prices. The loss incurred by the sale of these quantities outweighs the gains from a future monopoly position for the cartel. Stated in another way, the response of fringe sales to different inventory levels is too sluggish for cheating to work. For example, Salant (Ref. 6, p. 1085) shows that if demand is linear, then $q_1^c(t)$ is independent of the cartel's actual or announced reserve holdings. Thus, the cartel is unable to change fringe production by misrepresenting initial reserves. In general, continuous monitoring of output makes lies about reserves unprofitable and thus not worth worrying about. In Lewis and Schmalensee (Ref. 15), we assume that initial reserves are known but that the cartel output is unobservable. There, we find that it is profitable for the cartel to lie either by overstating or understating its sales for a finite early period.

4. Noncooperative Oligopoly

4.1. Assumptions and Notation

In this section, we briefly consider some properties of noncooperative oligopoly models of renewable resource markets. Aside from some analysis by Salant (Ref. 6, Appendix B), this appears to be the first reasonably systematic examination of such models.

Assume that there exist N "firms" (which may be ordinary firms, nations, partial cartels, or other reserve owners) capable of producing the resource. Let total initial reserves be I, let firm i's initial reserves be I_i, and let $S_i = I_i/I$, for all i. Similarly, let $q_i(t)$ be firm i's rate of production at time t. Let $Q(t)$ be the sum of the $q_i(t)$ and let $\sigma_i(t) = q_i(t)/Q(t)$. We assume that firm i's unit extraction cost is c_i, a constant. As before, $P(t) = P(Q(T))$ is the market price and r is the rate of discount.

Each firm is assumed to choose the time path of its output to maximize the present value of its net revenue, taking as given the outputs of the other firms. In equilibrium, no single firm can profitably alter its output strategy. It is clear that for firms, the discounted net marginal revenue at all t such that $q_i(t) > 0$ must be equal, and this common value must at least equal the discounted net price at all other times. If not, since the outputs of all other firms are taken as given, a shift in production policy would raise the present value of net revenue. If we let $a(P(t)) = -dP/dQ$, the equilibrium

$$P(t) - q_i(t)a(P(t)) - c_i = \lambda_i e^{rt} \qquad \text{if } q_i(t) > 0, \tag{34}$$

$$P(t) - c_i \leq \lambda_i e^{rt} \qquad \text{if } q_i(t) = 0, \tag{35}$$

and

$$I_i = \int_0^\infty q_i(t)\, dt \tag{36}$$

for $i = 1, \ldots, N$, along with the demand constraint $P(t) = P(Q(t))$. It is clear that λ_i measures the marginal (private) value to firm i of additional reserves at time zero.

4.2. Zero Extraction Cost

To focus on the implications of various elements of the general situation defined in Section 4.1, it will be useful to begin by assuming that all the c_i are zero. Extension to the case of equal positive c_i is not difficult. The simplest situation is then the case of constant elasticity of demand, case CE. If ε is the absolute value of demand elasticity, then condition (34) may be written

$$P(1 - \sigma_i/\varepsilon) = \lambda_i e^{rt}. \tag{37}$$

At some time t, let $\bar{\lambda}$ be the simple average of the λ_i of firms producing at a positive rate. Differentiation of (37), summation across i, and division then yields

$$\dot{P}/P = r = r(\bar{\lambda}/\lambda_i)(1 - \sigma_i/\varepsilon)/(1 - 1/N\varepsilon). \tag{38}$$

From this it is immediate that σ_i must be constant for all i in equilibrium, so that $\sigma_i = S_i$ for all i and t. It is clear from (37) that for such an equilibrium to exist, ε must exceed the largest of the S_i. All firms then produce forever, and price rises at the rate of interest. In this case, as was noted in Section 2, the function $P(t)$ is independent of market structure.

In fact, an explicit solution is easily obtained in this special case. If we choose units so that the demand curve can be written $Q = P^{-\varepsilon}$, we obtain

$$P(t) = (r\varepsilon I)^{-1/\varepsilon} e^{rt}, \tag{39}$$

$$\lambda_i = (r\varepsilon I)^{-1/\varepsilon}(1 - S_i/\varepsilon), \tag{40}$$

$$V_i = I_i (r\varepsilon I)^{-1/\varepsilon}, \tag{41}$$

where V_i is the present value of firm i's receipts. It is easy to show that $\lambda_i = \partial V_i / \partial I_i$, holding constant the reserve levels of other firms. Note that smaller firms (as measured by reserves) have larger λ_i and thus greater incentives for exploration. Thus, even though the price path over time in this model is independent of the distribution of reserves among producers, it seems likely that lower concentration would lead to more exploration in an expanded model.

Now let us consider case DE, in which demand elasticity falls with total output. The analysis is simplified by assuming that $S_i = 1/N$ for all i. Let $P(0) = F$, the choke price, be finite as before. By the symmetry of the problem, we can look for an equilibrium in which $\sigma_i = 1/N$ for all i and t. If T is the total time taken to exhaust the resource, the equilibrium conditions can be written

$$FE^{-ru} = P - (1/N)Q(P)a(P) = \mathrm{MR}(P), \qquad 0 \leq u \leq T, \tag{42}$$

$$\int_0^T Q(P(t))\, dt = I. \tag{43}$$

The assumption that demand elasticity rises with P can be seen to imply

$$a'Q < (\varepsilon + 1)/\varepsilon. \tag{44}$$

To establish the existence of a noncooperative equilibrium, we follow Salant's (Ref. 6) method of proof. Assume there exists a price $\hat{P}$ at which $N\varepsilon = 1$. For prices below $\hat{P}$, MR is negative, so that $P > \hat{P}$ at all times. Differentiation of (42) then establishes

$$\dot{P} = rP \frac{N\varepsilon - 1}{\varepsilon[(N + 1) - a'Q]}. \tag{45}$$

From (44), P is increasing for $P > \hat{P}$, so that Q is falling at all times. Further, (44) and (45) establish that $(\dot{P}/P) < r$ at all times and that in the limit as N increases, $(\dot{P}/P)$ approaches r.

Suppose one picks a $\tilde{P}_0$ and allows P to grow according to (45) until it equals F. As long as demand is smooth, the total production along this path, $\tilde{I}$, will be a continuous function of $\tilde{P}_0$. Clearly, $\tilde{I}(F) = 0$. As $\tilde{P}_0$ approaches $\hat{P}$ from above, I must increase without bound, along with the total time that the system spends with price arbitrarily near $\hat{P}$. It is easily seen that $\tilde{I}$ is monotone decreasing in $\tilde{P}_0$. Thus, there must exist a P_0 such that $\tilde{I}(P_0) = I$, and we have established the first part of the following proposition.

Proposition 4.1. If $c_i = 0$ and $S_i = 1/N$, for $i = 1, \ldots, N$, if demand is sufficiently smooth, and if there exists $\hat{P}$ such that $N\varepsilon(\hat{P}) = 1$, then there exists a unique noncooperative oligopoly equilibrium satisfying (42) and (43) in case DE. Price increases at less than the rate of interest. Increases in N (a) lower T, (b) raise P_0, (c) lower total profits, (d) raise the marginal value of reserves to each firm, λ, and (e) raise the discounted value of net welfare.

Let us first prove (a) and (b). We want to show that if $N' > N$, then $T' < T$ and $P_0' < P_0$, where primes are used to denote quantities associated with N', and unprimed quantities are associated with N. That is, increasing the number of firms with reserves held fixed leads to more rapid extraction. For $P > \hat{P}$, MR is easily seen to be increasing in P under DE. For fixed positive u (time until exhaustion), (42) then implies that $P' < P$, so that $Q' > Q$. Given the total output constraint (43), $Q' > Q$ implies that T' must be less than T. Since P is a monotone function of time, (43) can be transformed by a change of variable to

$$\int_{P_0}^{F} [Q(P)/\dot{P}(P)]\, dP = I. \tag{46}$$

Suppose that $P_0' = P_0$. From (44) and (45), $\dot{P}'(P) > \dot{P}(P)$ for $P \in [P_0, F)$, and $Q(P) = Q'(P)$. Thus, less total output is produced with N' than with N. The only way to satisfy the constraint is to have $P_0' < P_0$, and (a) and (b) of Proposition 4.1 are established.

Part (d) of Proposition 4.1 follows immediately from (a) and (34). With more competition, the incremental value of reserves rises for each firm. Heuristically, there are two opposed forces here: price tends to be lower, but marginal revenue tends to be closer to price. The second of these dominates. Exploration incentives are enhanced with other firms, even though prices are lower.

We can most readily establish (c) and (e) of Proposition 4.1 by writing the equilibrium output path as $Q(N, t)$ for fixed I. We suppose that demand is smooth enough that Q is differentiable in N. Then (42) can be rewritten as

$$P(Q(N, t)) - (1/N)Q(N, t)a(P(Q(N, t))) = \lambda e^{rt}. \tag{47}$$

Differentiation with respect to N for fixed t and substitution for aQ/N^2 yield

$$Q_N(N, t)aN^2(N + 1 - a'Q) = P - [N(\partial\lambda/\partial N) - \lambda]e^{rt}. \tag{48}$$

By (44) and $P > \hat{P}$, the quantity multiplying Q_N is positive. From (b) of Proposition 4.1, $Q_N(N, 0) > 0$, so that the right-hand side of (48) is initially positive. But since P grows at a rate less than r, (48) then implies that Q_N is initially positive, reaches zero in finite time, and is negative thereafter.

Total discounted profit for all firms can be written as

$$V = \int_0^T P(Q(N, t))Q(N, t)e^{-rt}\, dt. \tag{49}$$

Since $Q(N, T) = 0$ for any N, and $P(0) = F$ is finite, we have

$$\begin{aligned}\frac{\partial V}{\partial N} &= \int_0^T (P - aQ)Q_N(N, t)e^{-rt}\, dt \\ &= N\int_0^T \lambda Q_N(N, t)\, dt - (N - 1)\int_0^T (Pe^{-rt})Q_N(N, t)\, dt\end{aligned} \tag{50}$$

using (47). The first integral is zero from (43) and $Q(N, T) = 0$. Since P is always increasing at a rate less than r, (Pe^{-rt}) is everywhere decreasing. Since Q_N is positive for small t and negative for large t, integrating to zero overall, it follows that the second integral on the right side of (50) is positive. Thus, $\partial V/\partial N$ is negative, as was to be shown. This is an intuitive result; more competition lowers profits.

Finally, the sum of total discounted consumer's surplus and firm's rents is given by

$$W = \int_0^T \left[\int_0^{Q(N,t)} P(x)\, dx\right] e^{-rt}\, dt. \tag{51}$$

Since $Q(N, T) = 0$, differentiation yields

$$\frac{\partial W}{\partial N} = \int_0^P (Pe^{-rt})Q_N(N, t)\, dt. \tag{52}$$

which is positive by our preceding reasoning. We have thus completed the proof of Proposition 4.1 and have shown that under our assumption, increased competition improves performance. □

The asymmetric case, in which initial reserves and unit extraction costs are not equal, is considerably more complicated and is being investigated in a subsequent paper by the authors.†

Appendix A

Comparative static derivations for the first five rows of Table 1 are signed following a method developed by Hartwick (Ref. 14) for the (DE) demand specification; the other cases (CE) and (IE) are handled in a similar manner. Details of the other derivations in Table 1 are available on request from the authors.

Let us first consider the first two lines in Table 1. From equations (8), (10), and (18), we have

$$I^c = -\int_{q_1}^{q_0} \{1 - [1 - k(q_1)]\varepsilon(q)\}q/\dot{\bar{q}}^c\, dq, \tag{A1}$$

$$I^m = -\int_{q_1}^{q_0} [1 - k(q_1)\varepsilon(q)]q/\dot{\bar{q}}^c\, dq + -\int_0^{q_1} q/\dot{\bar{q}}^m\, dq. \tag{A2}$$

Total differentiation of equations (A1) and (A2) yields a set of equations that can be represented in matrix form as

$$\begin{bmatrix} \alpha_{11} & \alpha_{12} \\ \alpha_{21} & \alpha_{22} \end{bmatrix}\begin{bmatrix} dq_0 \\ dq_1 \end{bmatrix} = \begin{bmatrix} dI^c \\ dI^m \end{bmatrix}, \tag{A3}$$

where

$$\alpha_{11} = -\{1 - [1 - k(q_0)]\varepsilon(q_0)q_0\}/\dot{\bar{q}}^c(q_0) > 0,$$

$$\alpha_{12} = -\int_{q_1}^{q_0} \varepsilon(q)q\left(\frac{dk}{dq_1}\right)\Big/\dot{\bar{q}}^c\, dq < 0,$$

$$\alpha_{21} = -[1 - k(q_1)]\varepsilon(q_1)q_1/\dot{\bar{q}}^c(q_1) > 0,$$

$$\alpha_{22} = \int_{q_1}^{q_0} \frac{dk}{dq_1}\frac{\varepsilon(q)}{\dot{\bar{q}}^c}\, dq - \frac{q_1}{\dot{\bar{q}}^m(q_1)} > 0,$$

$$\frac{dk}{dq_1} = \frac{\varepsilon'(q_1)}{\varepsilon^2(q_1)} < 0.$$

† In the case $c_i = 0$, $i = 1, 2, \ldots, N$, and $S_i < S_j$ for some pair of firms we find that if a noncooperative equilibrium exists, then (a) prices rise at less than the rate of interest, (b) firm i exhausts reserves and ceases production as q_1 falls to zero before firm j, and (c) the initial price P_0 is higher the more unequal the initial distribution of reserves. If in addition, $0 < c_i < c_j$ for some i and j, then the total discounted cost of producing the equilibrium total output path, $Q(t)$ is not minimized.

Repeated application of Cramer's rule yields

$$dq_0/dI^c > 0, \tag{A4}$$

$$dq_0/dI^m > 0, \tag{A5}$$

$$dq_1/dI^c > 0, \tag{A6}$$

$$dq_1/dI^m > 0, \tag{A7}$$

$$dq_0/dS = dq_0/dI^m - dq_0/dI^c < 0, \tag{A8}$$

$$dq_1/dS = dq_1/dI^m - dq_1/dI^c > 0, \tag{A9}$$

where a comparison of relative magnitudes in equations in (A4) and (A5), and (A6) and (A7) are required to sign dq_0/dS and dq_1/dS. Equations (A4), (A5), (A8), and the condition $p' < 0$ imply the signs appearing in the second row of Table 1. From (1) and (3) we have that whenever $q^c(t)$, and $q^m(t) > 0$,

$$\text{MR}_p^m = kp_0. \tag{A10}$$

Differentiating (A10) first with respect to I^m and then with respect to I^c yields

$$\partial\text{MR}_0^m/\partial I^m = (dk/dq_1)(\partial q_1/\partial I^m)p_0 + k\partial p_0/\partial I^m < 0, \tag{A11}$$

$$\partial\text{MR}_0^m/\partial I^c = (dk/dq_1)(\partial q_1/\partial I^c)p_0 + k\partial p_0/\partial I^c < 0, \tag{A12}$$

$$\partial\text{MR}^m/\partial S = \partial\text{MR}^m/\partial I^m - \partial\text{MR}^m/\partial I^c < 0, \tag{A13}$$

where a comparison of relative magnitudes in (A11) and (A12) is required to sign (A13).

A similar approach is used for lines 3–5 in Table 1. The functions $W^m(I^m, I^c)$ and $W^c(I^m, I^c)$ are defined by

$$W^m = \int_0^{T_2} e^{-rt}p(q^e + q^m)q^m\,dt, \tag{A14}$$

$$W^c = \int_0^{T_1} e^{-rt}p(q^c + q^m)q^c\,dt, \tag{A15}$$

where T_1 and T_2 are the time horizons over which fringe suppliers and the cartel operate, respectively, with $T_2 > T_1$.

Straightforward differentiation of W^m with respect to I^m yields

$$\partial W^m/\partial I^m = \int_0^{T_2} e^{-rt}\left[\frac{p\,dq^m}{dI^m} + \frac{dp}{dq}\left(\frac{dq^m}{dI^m} + \frac{dq^c}{dI^m}\right)q^m\right]dt + \frac{\partial T_2}{\partial I^m}e^{-rT_2}p(T_2)q^m(T_2). \tag{A16}$$

The second term on the right-hand side of (A16) is zero since $q^m(T_2) = 0$. Rearranging (A16), we obtain

$$\frac{\partial W^m}{\partial I^m} = \int_0^{T_2} e^{-rt}\left[\mathrm{MR}^m \frac{dq^m}{dI^m} + (\mathrm{MR}^m - p)\frac{dq^c}{dI^m}\right] dt. \tag{A17}$$

To preserve our inventory constraints we have, from (2),

$$\frac{\partial I^c}{\partial I^m} = 0 = \int_0^{T_1} \frac{\partial q^c}{\partial I^m}\, dt + \frac{\partial T_1}{\partial I^m} q^m(T_1), \tag{A18}$$

$$\frac{\partial I^m}{\partial I^m} = 1 - \int_0^{T_2} \frac{\partial q^m}{\partial I^m}\, dt. \tag{A19}$$

Note that $q^m(T_1) = 0$ in order to satisfy (1) and (3). Using (3), (13), (A18), and (A19), we obtain

$$\begin{aligned}\frac{\partial W^m}{\partial I^m} &= \mathrm{MR}_0^m \int_0^{T_2} \frac{dq^m}{dI^m}\, dt + (\mathrm{MR}_0^m - p_0) \int_0^{T_1} \frac{dq^c}{dI^m}\, dt \\ &= \mathrm{MR}_0^m > 0.\end{aligned} \tag{A20}$$

Straightforward but somewhat tedious application of this same method can be shown to yield the following derivatives with the indicated signs:

$$\frac{\partial W^m}{\partial I^c} = \mathrm{MR}_0^m - p_0 < 0, \tag{A21}$$

$$\frac{\partial W^m}{\partial S} = \frac{\partial W^m}{\partial I^m} - \frac{\partial W^m}{\partial I^c} = p^0 > 0, \tag{A22}$$

$$\frac{\partial W^c}{\partial I^m} = \int_0^{T_1} e^{-rt}\left[\frac{dp}{dq} q^c \frac{dq}{dI^m}\right] dt < 0, \tag{A23}$$

$$\frac{\partial W^c}{\partial I^c} = p_0 + \int_0^{T_1} e^{-rt}\left[\frac{dp}{dq} q^c \frac{dq}{dI^c}\right] dt > 0, \tag{A24}$$

$$\frac{\partial W^c}{\partial S} = \frac{\partial W^c}{\partial I^m} - \frac{\partial W^c}{\partial I^c} < 0, \tag{A25}$$

$$\frac{\partial (W^m + W^c)}{\partial S} = \frac{\partial W^m}{\partial S} + \frac{\partial W^c}{\partial S} > 0. \tag{A26}$$

Appendix B

The condition

$$q_1^m(T) \gtreqless q_2^m(T) \qquad \text{as } \bar{I}^m \lesseqgtr I^m(0) \tag{B1}$$

for (DE) specification is derived as follows.

a. Let $p(t, \bar{I}^m, I^c)$ and $\mathrm{MR}^m(t, \bar{I}^m, I^c)$ be the Nash equilibrium price and marginal revenue paths. From Table 1, $\partial P(0)/\partial \bar{I}^m < 0$, implying an initial downward shift in prices as $\bar{I}^m$ increases. In addition, $\mathrm{MR}_0^m(0) = e^{-rT}F$, where F is equal to the choke price and T is the time it takes to exhaust total inventories. Since $\partial \mathrm{MR}^m(0)/\partial \bar{I}^m = -re^{-rT}F(\partial T/\partial \bar{I}^m) < 0$, we conclude that $\partial T/\partial I^m > 0$, which combined with our earlier argument can be shown to imply that $\partial p(t)/\partial \bar{I}^m < 0\ (\forall t)$, or $\partial q(t)/\partial \bar{I}^m > 0$. Since $q^m(t) = q(t)$ for $t > t_1$, we have

$$\frac{\partial q^m(t)}{\partial \bar{I}^m} > 0, \qquad t > t_1. \tag{B2}$$

b. We now wish to sign $\partial q^m(t)/\partial \bar{I}^m$ for $t \geq t_1$, which we will do by examining the sign of $\partial q^c(t)/\partial \bar{I}^m$. Since $\partial q(t)/\partial \bar{I}^m > 0$, we have

$$\operatorname{sign}\frac{\partial q^c(t)}{\partial \bar{I}^m} = \operatorname{sign}\left[\frac{\partial q^c(t)}{\partial q(t)}\right]\left[\frac{\partial q(t)}{\partial \bar{I}^m}\right] = \operatorname{sign}\frac{\partial q^c}{\partial \bar{q}(t)}. \tag{B3}$$

Given (DE′) and (1) and (3)

$$q^m(t) = \frac{p(t) - \mathrm{MR}^m(t)}{(\alpha + 1)q(t)^\alpha} = \frac{e^{-rt}[p(0) - \mathrm{MR}^m(0)]}{(\alpha + 1)q(t)^\alpha}. \tag{B4}$$

Straightforward differentiation of $q^c(t)$ with respect to $q(t)$, substituting for $q^m(t)$ from (B4) yields

$$\frac{dq^c(t)}{dq(t)} = 1 - e^{-rt} \times \left(\frac{\dfrac{d}{dq(0)}[p(0) - \mathrm{MR}^m(0)]e^{-rt}\left\{\dfrac{q(t)^\alpha}{q(0)} - \dfrac{\alpha}{q(t)}[p(0) - \mathrm{MR}^m(0)]\right\}}{(\alpha + 1)q(t)^\alpha}\right). \tag{B5}$$

From (B5) we can establish that

$$\frac{d}{dq(0)}[P(0) - \mathrm{MR}^m(0)] > 0. \tag{B6}$$

To see this, suppose that $[d/dq(0)]\,[p(0) - \mathrm{MR}^m(0)] < 0$. Then if $\alpha \geq 0$, (B5) implies that $\partial q^c(t)/\partial \bar{I}^m > 0\ (\forall t)$, which violates (3). Differentiating (B5) with respect to time, we obtain

$$\frac{d}{dt}\left[\frac{dq^e(t)}{dq(t)}\right] = r\left\{\frac{dq^e}{dq(t)} - 1\right\} - e^{-rt} \times \left\{\frac{-re^{-rt}\dfrac{d}{dq(0)}(P(0) - \mathrm{MR}^m(0)) \times [q(0)^{-\alpha} + (\alpha + 1)]q^{-(\alpha+2)}\dfrac{dq}{dt}[P(0) - \mathrm{MR}^m(0)]}{\alpha + 1}\right\}. \tag{B7}$$

If $\alpha < 0$ and $dq^c(t)/dq(t) \leq 0$ for some t', then (B7) implies $dq^c(t)/dq(t) < 0$ for all $t \geq t'$. Clearly, $\partial q^c(t)/\partial q(t) \leq 0$ for some $t' \in (0, t_1)$ (otherwise (3) is violated), but this implies that $\partial q^c(t_1)/\partial q(t_1) < 0$, which is not an allowable variation since $q^c(t_1) = 0$. Consequently, (B6) must hold.

Given (B6) we can now establish the signs of $\partial q^c(t)/\partial q(t)$ and $\partial q^m(t)/\partial q(t)$. For $\alpha < 0$ it follows from (B5) that $\partial q^c(t)/\partial q(t) < 1$, which implies that $\partial q^m/\partial q(t) > 0$ for all t. For $\alpha \geq 0$ we also have that $\partial q^c(t)/\partial q(t) < 1$ for $t \in [0, t_1]$, for if $\partial q^c(t)/\partial q(t) > 1$ at some $t' < t_1$, then (B7) implies that $\partial q^m(t)/\partial q(t) < 0$ for all $t' < t \leq t_1$. But given (B2), $q^m(t)$ is discontinuous at t_1, implying that $\mathrm{MR}^m(t_1^-) > \mathrm{MR}^m(t_1)$ which violates (3). Thus we have established

$$\frac{\partial q^m(t)}{\partial q(t)} > 0 \qquad (\forall t). \tag{B8}$$

Condition (B1) now follows immediately from (B8) and the fact that $\partial q(t)/\partial \bar{I}^m > 0$.

References

1. Scherer, F. M., *Industrial Market Structure and Economic Performance,* Rand-McNalley, Chicago, 1970.
2. Clark, C., *Restricted Entry to Common Property Fishery Resources: A Game Theoretic Analysis,* Technical Report No. 78-9, The Institute of Applied Mathematics and Statistics, University of British Columbia, 1978.
3. Levhari, D. and Mirman, L., *The Great Fish War: An Example Using Dynamic Cournot–Nash Solution,* Department of Economics, University of Illinois, Urbana, 1977.
4. Stiglitz, J. E., Monopoly and the rate of extraction of exhaustible resources, *American Economic Review,* Vol. 66, pp. 655–661, 1976.
5. Hotelling, H., The economics of exhaustible resources, *Journal of Political Economy,* Vol. 39, pp. 137–175, 1931.
6. Salant, S. W., Exhaustible resources and industrial structure; A Nash–Cournot approach to the world oil market, *Journal of Political Economy,* Vol. 84, pp. 1079–1093, 1976.
7. Pindyck, R. S., Cartel pricing and the structure of the world bauxite market, *Bell Journal of Economics,* Vol. 8, pp. 343–360, 1977.
8. Pindyck, R. S., Gains to producers from the cartelization of exhaustible resources, *Review of Economics and Statistics,* Vol. 60, pp. 238–251, 1978.
9. Eckbo, P. L., *The Future of World Oil,* Ballinger, Cambridge, Mass., 1976.
10. Hynilicza, E. and Pindyck, R., Pricing policies for a two-part exhaustible resource cartel: The case of OPEC, *European Economic Review,* Vol. 8, pp. 139–154, 1976.
11. Kosbud, R. F. and Stokes, H. H., *Economic Analysis of OPEC Using a Markov Model,* Department of Economics, University of Illinois, Chicago, 1977.
12. Pakravan, K., *The Theory of Exhaustible Resources and Market Organization with an Application of Oil and OPEC,* Department of Economics, University of Chicago, Ph.D. Dissertation, 1976.
13. Schmalensee, R., Is more competition necessarily good? *Industrial Organization Review,* Vol. 4, pp. 120–121, 1976.

14. HARTWICK, J., Exploitation of many deposits of an exhaustible resource, *Econometrics*, Vol. 46, pp. 201–219, 1977.
15. LEWIS, T. R. and SCHMALENSEE, R. L., *Cartel Deception in Nonrenewable Resource Markets*, Working Paper No. 1010–78, Massachusetts Institute of Technology, July 1978.

Supplementary References

NICHOLS, A. L. and ZECKHAUSER, R., Stockpiling strategies and cartel prices, *Bell Journal of Economics*, Vol. 8, pp. 66–96, 1977.

RADER, T., *Theory of Microeconomics*, Academic Press, New York, 1972.

SCHMALENSEE, R., Resource exploitation theory and the behavior of the oil cartel, *European Economic Review*, Vol. 7, pp. 257–279, 1976.

9

Optimum Extraction of an Exhaustible Resource: A Mathematical Analysis

PAN-TAI LIU

1. Introduction

In this paper, we discuss optimum extraction of an exhaustible resource. The problem is essentially the same as in (Ref. 1), but we shall study it in greater detail and in a somewhat more general context.

The initial amount of reserve is unknown, but some *a priori* knowledge about it is given by a probability distribution function. A general profit function and cost function associated with extraction are assumed. The firm's objective is to maximize the expected value of discounted utility over an infinite horizon.

In Section 2, we consider one-firm exploitation. We analyze the optimum solution by invoking the maximum principle. Moreover, we propose a transversality condition, based on which the final level of production can be determined. A procedure for actually obtaining the optimum extraction rate is also outlined.

In Section 3, we use dynamic programming to derive a first-order differential equation for an optimal closed-loop control (i.e., extraction rate). While this equation does not yield to a simple analytic solution in general, it does shed some light on the effect of uncertainty.

Finally, in Section 4, we consider n-firm exploitation of a common resource. We seek a set of Pareto-optimal controls and derive an equilibrium condition, which shows the interconnection among the n firms under such optimality.

PAN-TAI LIU • Mathematics Department, University of Rhode Island, Kingston, Rhode Island.

The formulation and treatment of the extraction problem in this paper admittedly neglect certain economic aspects of the problem. Nevertheless, from a control-theoretic viewpoint, the purpose is to study the basic mathematical problem with a view to proper characterization of the optimal solution. It is hoped that such a mathematical approach will be useful for elucidation of some economic theory underlying the problem of exploitation of exhaustible resources.

2. Exploitation by One Firm

The dynamic equation for case is simply

$$\frac{dx}{dt} = \dot{x} = q, \qquad x(0) = 0, \tag{1}$$

where $0 \leq x \leq x_m$ and $q \geq 0$. The state variable $x = x(t)$ is the accumulated amount of extraction up to t, and the control variable $q = q_t$ is the extraction rate.

The initial amount of reserve Z is assumed to be a continuous random variable with a probability distribution function $1 - F(x)$, $0 \leq x \leq x_m$, i.e., $F(x) = P\{Z > x\}$. The function $F(x)$ is continuously differentiable, with $F(0) = 1$, $F(x_m) = 0$, and $F'(x) < 0$, $0 < x < x_m$. Therefore x_m is the maximum possible amount of the initial reserve.

The firm seeks to determine an optimal extraction policy q_t^* to maximize the expected discounted utility over an infinite horizon:

$$J = E\int_0^\infty e^{-\delta t} U(\Pi_t)\, dt, \tag{2}$$

where $\delta > 0$ is a discount factor, U the utility function, Π_t the instantaneous revenue of the firm. The utility function U satisfies the conditions $U(0) = 0$, $U'(\Pi) > 0$, and $U''(\Pi) < 0$ for all $\Pi > 0$ and $\lim_{\Pi\to 0} U'(\Pi) = \infty$.

The revenue can be written $\Pi = P(q) - C(x, q)$, where $P(q)$ is the profit function and $C(x, q)$ the cost function associated with extraction. If the unit price of the output is a constant p, then $P(q) = pq$. In general, we assume that $P(0) = 0$, $P'(0) > 0$, and $P''(q) \leq 0$ for all $q \geq 0$. Also, $C(x, q)$ satisfies the conditions $C(x, 0) = 0$ for all $x \geq 0$ and $C_x \geq 0$, $C_q \geq 0$, $C_{qq} \geq 0$, and $C_{qx} \geq 0$ for all $x \geq 0$ and $q \geq 0$.

This is basically a stochastic control problem. The actual outcome of exploitation is a stopped process $z(t, \omega)$ where $\omega \in \Omega$ and Ω is a probability space. This means that $z(t, \omega) = x(t)$ whenever $x(t) \leq Z$ and $z(t, \omega)$ stops at the instant when $x(t) = Z$. After this time, extraction must stop.

However, the objective functional (2) can be shown to be equivalent to the following†

$$J = \int_0^\infty e^{-\delta t} U(\Pi_t) F(x(t))\, dt. \tag{3}$$

The firm can then plan the extraction policy over an infinite horizon as if extraction would continue forever (while in reality it may terminate at any time because of depletion, and a positive q at any t is the actual extraction rate only if depletion has not occurred up to t). As a consequence of replacing (2) by (3), the problem can be treated as one of deterministic optimal control.

We now apply the maximum principle and write the Hamiltonian

$$H = \phi q + U(\Pi) F(x), \tag{4}$$

where $\phi = \phi(t)$ satisfies

$$\dot{\phi} = \delta\phi - U(\Pi_t^*) F'(x^*(t)) + U'(\Pi_t^*) C_x(x^*(t),\ q_t^*) F(x^*(t)). \tag{5}$$

We then have, for all $t \geq 0$,

$$H(q_t^*, x^*(t), \phi) = \max_{q \geq 0} H(q, x^*(t), \phi). \tag{6}$$

Let us define $S = \{x \mid 0 \leq x < x_m \text{ and } P'(0) - C_q(x, 0) > 0\}$. To maximize H, we have

$$q^* = 0, \qquad x^* \notin S,$$

and

$$\frac{\partial H}{\partial q^*} = 0, \qquad x^* \in S,$$

or, equivalently,

$$\phi = -U'(\Pi_t^*)\,(P'(q^*) - C_q(x^*, q^*)) F(x^*). \tag{7}$$

For every $x^* \in S$, the partial derivative of the right-hand side of (7) with respect to q is positive, since

$$-U''(\Pi^*)(P'(q^*) - C_q(x^*, q^*))^2 F(x^*) \\ - U'(\Pi^*)(P''(q^*) - C_{qq}(x^*, q^*)) F(x^*) > 0. \tag{8}$$

From (7), we can now synthesize q^*, i.e.,

$$\begin{aligned} q^* &= q^*(x^*, \phi), \qquad && x^* \in S, \\ &= 0, && x^* \notin S. \end{aligned} \tag{9}$$

† This actually requires that $E(\tau(\omega)) < \infty$, where $\tau(\omega)$ is the random time at which $x(t) = Z$. We assume that this requirement is satisfied here.

Substituting (9) into (1) and (5), we have a pair of differential equations for x^* and ϕ. In order to solve these equations, we need some conditions on the limiting behavior of x^* and ϕ as $t \to \infty$. The conditions can be stated as follows.

Proposition 2.1. Let $x^*(t)$ and $\phi(t)$ be solutions to (1) and (5) with q^* given by (9). In order for $x^*(t)$ to be optimal, the following must be true:

I. $\lim_{t\to\infty} \phi(t) = 0$ if $\lim_{t\to\infty} x^*(t) = \hat{x} < x_m$;

II. $\lim_{t\to\infty} \phi(t) = \alpha \leq 0$ if $\lim_{t\to\infty} x^*(t) = x_m$.

This proposition is a special case of Proposition 4.2, whose proof is outlined in the appendix.

Let us discuss the two cases separately.

Case I. Since $F(\hat{x}) > 0$, $\lim_{t\to\infty} q_t^* = 0$, and $\lim_{t\to\infty} U'(\Pi_t^*) = \infty$, we must have, from (7),

$$P'(0) = C_q(\hat{x}, 0), \tag{10}$$

from which $\hat{x}$ can be determined (see Fig. 1). Obviously, in this case, $S = [0, \hat{x})$ and $x^*(t) \in S$ for all $t \geq 0$.

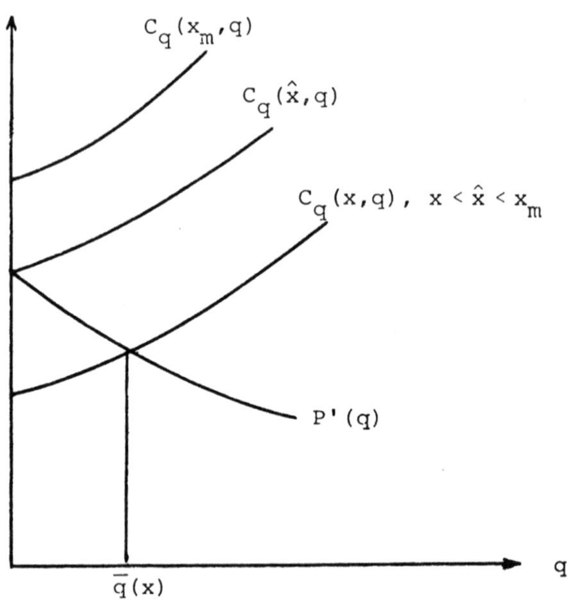

Figure 1. Case 1: $S = [0, \hat{x})$.

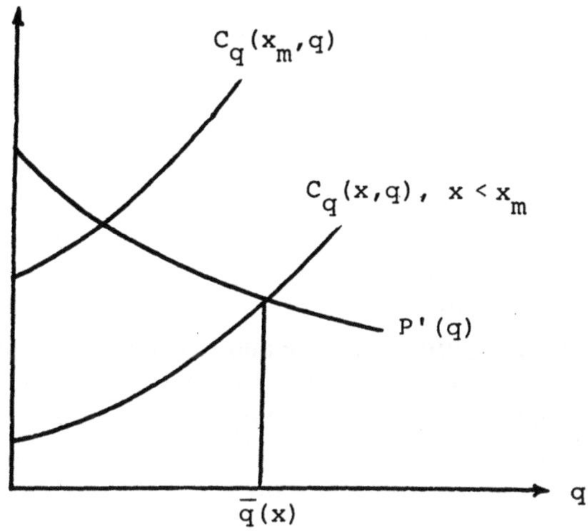

Figure 2. Case : $S = [0, x_m)$.

Knowing the limits of $x^*(t)$ and $\phi(t)$, we can solve (1) and (5) (with q^* given by (9)) as follows.†

(i) Assume "initial values" for x^* and ϕ, namely, $x^* = \hat{x} - \varepsilon_1$ and $\phi = -\varepsilon_2$, where $\varepsilon_i (i = 1, 2)$ is an arbitrarily small positive number.

(ii) Solve (1) and (5) by integrating backwards and obtain $x^* = \bar{x}^*(t, \varepsilon_1, \varepsilon_2)$ and $\phi = \bar{\phi}(t, \varepsilon_1, \varepsilon_2)$.

(iii) Take the limits as $\varepsilon_1 \to 0$ and $\varepsilon_2 \to 0$ and obtain the desired solutions, i.e.,

$$x^*(t) = \lim_{\varepsilon_1 \to 0, \varepsilon_2 \to 0} \bar{x}^*(t, \varepsilon_1, \varepsilon_2)$$

and

$$\phi(t) = \lim_{\varepsilon_1 \to 0, \varepsilon_2 \to 0} \bar{\phi}(t, \varepsilon_1, \varepsilon_2).$$

Case II. Since $F(x_m) = 0$ and $\lim_{t \to \infty} U'(\Pi_t^*) = \infty$, we must have $P'(0) - C_q(x_m, 0) \geq 0$. In this case $S = [0, x_m)$ and again $x^*(t) \in S$, $t \geq 0$ (see Fig. 2).

We can determine α by setting $\dot{\phi} = 0$ in (5), and we obtain

$$\alpha = \frac{1}{\delta} \lim_{x \to x_m, q \to 0} [U(\Pi)F'(x) - U'(\Pi)C_x(x, q)F(x)]. \tag{11}$$

We then proceed to solve (1) and (5) in a way identical to Case I.

† This is the same method that is used by Chang in Part I of this volume.

The "initial values" for x^* and ϕ in step (i) are $x_m - \varepsilon_1$ and $\alpha - \varepsilon_2$ respectively.

After $x^*(t)$ and $\phi(t)$ are obtained, q_t^* can finally be determined from (9).

Let us recapitulate the significance of the two cases regarding the final outcome of exploitation. In Case I, it follows that $P\{x^*(t) = Z$ for some $t \geq 0\} = 1$ and therefore $z^*(t, \omega)$ stops at some finite time. This means that the resource is depleted with probability one. In Case II, we have $P\{x^*(t) < Z$, for all $t \geq 0\} = P\{\hat{x} < Z\} = F(\hat{x})$, which is the probability that the stopping time is infinite and hence that the resource will not be depleted. This happens when the marginal profit is not high enough, especially for small q, or when the marginal cost of extraction becomes too high after a certain level of production is reached, and it is not lucrative for the firm to exploit any further.

3. Dynamic Programming Approach

In this section, we use the method of dynamic programming to seek a closed-loop optimal control $q^*(x)$.

To begin with, let us consider a static problem of maximizing $U(\Pi)$. The first-order condition dictates that

$$\frac{\partial U}{\partial q} = (P'(q) - C_q(x, q))U'(\Pi) = 0. \tag{12}$$

Since $U' > 0$, we must have

$$P'(q) - C_q(x, q) = 0 \tag{13}$$

from which $q = \bar{q}(x)$ can be determined (see Fig. 1).

According to the result in Section 2, $x^*(t) \in S$ for all $t \geq 0$. Thus for every $x \in S$ and $x^*(t) = x$ we define the optimum value function

$$\mathcal{V}(x, t) = \max_{q \geq 0} \int_t^\infty e^{-\delta s} U(\Pi_s) F(x(s))\, ds. \tag{14}$$

Using discounted dynamic programming, we can show that $\mathcal{V}(x, t) = V(x)e^{-\delta t}$, where $V(x)$ satisfies

$$\max_{q \geq 0} [V'(x)q - \delta V(x) + U(\Pi)F(x)] = V'(x)q^* - \delta V(x) + U(\Pi^*)F(x) = 0 \tag{15}$$

for every $x \in S$.

Suppose that $q^* = 0$ at $x = \alpha \in S$. From the second part of (15), it then follows that $V(\alpha) = 0$. This means that $q^*(x) = 0$ for all $x \in S$ such that $x \geq \alpha$ since the process of exploitation must stop at $x = \alpha$. But this contradicts the conclusion in Section 2 that either $\lim_{t\to\infty} x^*(t) = \hat{x}$ or $\lim_{t\to\infty} x^*(t) = x_m$. Hence $q^*(x) > 0$ for all $x \in S$.

We can now differentiate with respect to q in (15) and obtain, for $x \in S$,

$$V'(x) = -(P'(q^*) - C_q(x, q^*))U'(\Pi^*)F(x). \tag{16}$$

It can be shown by using the second derivative test that (16) is indeed the necessary condition for the maximum in (15).

Given that $q^*(x) > 0$, V must be a decreasing function of x and $V'(x) \leq 0$ for all $x \in S$. Since $U' > 0$ and $F > 0$, from (16), we obtain

$$P'(q^*) - C_q(x, q^*) \geq 0, \tag{17}$$

which implies that $q^*(x) \leq \bar{q}(x)$ (see Figs. 1 and 2), an obvious fact that the optimum extraction rate is lower in the dynamic case than in the static case in order to conserve for future consumption. Now combining (16) with the second part of (15) yields

$$V(x) = \frac{1}{\delta}[U(\Pi^*) - (P'(q^*) - C_q(x, q^*))q^*U'(\Pi^*)]F(x). \tag{18}$$

The function V given by (18) is always positive, as it should be, for $x \in S$. This can be shown by applying the mean value theorem, namely,

$$U(\Pi^*) = (P'(\tilde{q}) - C_q(x, \tilde{q}))U'(\tilde{\Pi})q^*, \tag{19}$$

where $0 \leq \tilde{q} < q^*$, and $\tilde{\Pi} = P(\tilde{q}) - C(x, \tilde{q})$. But $\Pi^* > \hat{\Pi}$, $U'(\hat{\Pi}) > U'(\Pi^*)$, and $P'(\tilde{q}) - C_q(x, \tilde{q}) > P'(q^*) - C_q(x, q^*)$, $x \in S$. We therefore obtain, from (18),

$$U(\Pi^*) > (P'(q^*) - C_q(x, q^*))U'(\Pi^*)q^*, \tag{20}$$

and $V(x) > 0$.

Next we differentiate $V(x)$ in (16) and substitute $V'(x)$ into (16). After some lengthy manipulation, we obtain, for $x \in S$,

$$\begin{aligned}
&[U'(\Pi^*)(P''(q^*) - C_{qq}(x, q^*)) - U''(\Pi^*)(P'(q^*) - C_q(x, q^*))]q^*q^{*\prime} \\
&\quad = [U'(\Pi^*)(P'(q^*) - C_q(x, q^*))q^* - U(\Pi^*)]\frac{F'(x)}{F(x)} \\
&\qquad + [C_{qx}(x, q^*)q^* - C_x(x, q^*)]U'(\Pi^*) \\
&\qquad - [\delta U'(\Pi^*) + U''(\Pi^*)C_x(x, q^*)q^*][P'(q^*) - C_q(x, q^*)]
\end{aligned} \tag{21}$$

where $q^{*\prime} = dq^*/dx$.

Equation (21) can be considered a first-order differential equation for q^*. A unique solution to the equation satisfying $0 < q^*(x) \leq \bar{q}(x)$ and

$\lim_{x \to \hat{x}} q^*(x) = 0$ when $S = [0, \hat{x})$ (or $\lim_{x \to x_m} q^*(x) = 0$ when $S = [0, x_m)$), if it exists, would be the optimal control.

While (19) does not yield to any simple analytic solution, it may still be easier to solve (19) in S rather than (1) and (5) over an infinite horizon. Moreover, it does shed some light on the effect of uncertainty. We explain this in the following.

In (20), only the first term on the right-hand side involves $F(x)$. Suppose that there is no uncertainty; then $F(x) = 1$ and $F'(x) = 0$ for $x \in (0, x_m)$, i.e., the initial amount of reserve is known to be exactly x_m. Let $q_0^*(x)$ be the solution to (21) in such a case, satisfying $0 < q_0^* \leq \bar{q}(x)$ and $\lim_{x \to \hat{x}} q_0^*(x) = 0$ (or $\lim_{x \to x_m} q_0^*(x) = 0$).†

Since the expression in brackets on the left-hand side of (21) is positive and the first term on the right-hand side is also positive, it is easy to see that for x sufficiently close to $\hat{x}$ (or x_m) we have $0 > q_0^{*\prime}(x) > q^{*\prime}(x)$ and hence $q_0^*(x) > q^*(x) > 0$. Furthermore, it is not possible for $q_0^*(x)$ to become less than $q^*(x)$ for any $x \in S$ because, if it were, then $q_0^* = q^*$ at some $\beta \in S$. This would require that $q^{*\prime} < q_0^{*\prime}$ at $x = \beta$. But this contradicts (21), from which we always have $q^{*\prime} > q_0^{*\prime}$ whenever $q^* = q_0^*$.

Therefore in the uncertainty case, the firm always exploits at a slower rate for fear of depletion in finite time. The difference between q^* and q_0^* depends on $-F'/F$. This latter function is the conditional probability density function of depletion at any instant, given that the accumulated amount of extraction up to the instant is x and depletion has not occurred up to that instant. For example, if $-F'/F \approx 0$, $x \in [\gamma, \hat{x})$ (or $x \in [\gamma, x_m)$) where $\gamma \in S$, then we have $q^* \approx q_0^*$. This means that if no depletion occurs before the level of production reaches γ, the firm is rather confident that the actual initial reserve is at least nearly $\hat{x}$ (or x_m) and extracts at almost the same rate as in the certainty case. With appropriate choices of some particular functions for U and Π, it is possible to determine more specifically how q^* is affected by $-F'/F$.

4. Pareto Optimum: Common Property Exploitation

We now consider n firms exploiting on a common property. Let x_i $(i = 1, 2, \ldots, n)$ be the accumulated amount of extraction and q_i $(i = 1, 2, \ldots, n)$ the extraction rate of each firm. The state equations are

$$\dot{x}_i = q_i, \qquad i = 1, 2, \ldots, n, \tag{22}$$

where $q_i \geq 0$ and $0 \leq \sum_{i=1}^{n} x_i = X \leq X_m$. As in the scalar case, Z is

† Note that whether $S = [0, \hat{x})$ or $S = [0, x_m)$ is independent of $F(x)$.

assumed to be a continuous random variable with the same probability distribution function $1 - F(X)$, i.e., $P\{Z > X\} = F(X)$.

Each firm has its own objective functional,

$$J_i = \int_0^\infty e^{-\delta t} U_i(\Pi_i) F(X)\, dt, \qquad i = 1, 2, \ldots, n, \tag{23}$$

where

$$\Pi_i = P_i(q_i) - C_i(X, q_i), \tag{24}$$

and P_i and C_i are the profit and cost functions of the ith firm respectively. They satisfy the same properties as P and C in Section 2, i.e., $P_i(0) = 0$, $P_i'(0) > 0$, $P_i'' < 0$ for all $q_i \geq 0$, and $C_i(X, 0) = 0$, $\partial C_i/\partial X \geq 0$, $\partial C_i/\partial q_i \geq 0$, $\partial^2 C_i/\partial q_i^2 \geq 0$, and $\partial^2 C_i/\partial X \partial q_i \geq 0$ for all $q_i \geq 0$ and $X \geq 0$.

A Pareto optimum is often the solution to such an economic system. This means that for some $\gamma_i > 0$, $i = 1, 2, \ldots, n$, we seek to determine a set of n optimal controls q_i^*, $i = 1, 2, \ldots, n$ in order to maximize $J = \sum_{i=1}^n \gamma_i J_i$.

This is an optimal control problem with bounded states: $x_i \geq 0$, $i = 1, 2, \ldots, n$, and $\sum_{i=1}^n x_i \leq X_m$. For Pareto optimality, we again invoke the maximum principle with the Hamiltonian written as

$$H = \sum_{i=1}^n \phi_i q_i + F(X) \sum_{i=1}^n \gamma_i U_i(\Pi_i), \tag{25}$$

where

$$\dot{\phi}_i = \delta\phi_i - F'(X^*) \sum_{i=1}^n \gamma_i U_i(\Pi_i^*) + F(X^*) \sum_{i=1}^n \gamma_i U_i'(\Pi_i^*) \frac{\partial C_i}{\partial X}. \tag{26}$$

Let us define $S_i = \left\{ X \mid 0 \leq X < X_m \text{ and } P_i'(0) - \frac{\partial C_i}{\partial q_i}(X, 0) > 0 \right\}$.

To maximize H, we have

$$q_i^* = 0, \qquad X^* \notin S_i,$$

and

$$\frac{\partial H}{\partial q_i^*} = 0, \qquad X^* \in S_i,$$

or, equivalently,

$$\phi_i = -\gamma_i U_i'(\Pi_i^*) \left(P_i'(q_i^*) - \frac{\partial C_i}{\partial q_i}(X^*, q_i^*) \right) F(X^*). \tag{27}$$

From (27), we can again synthesize q_i^*, i.e.,

$$\begin{aligned} q_i^* &= q_i^*(X, \phi_i), \qquad X \in S_i, \\ &= 0, \qquad\qquad X \notin S_i, \quad i = 1, 2, \ldots, n. \end{aligned} \tag{28}$$

Substituting (28) into (22) and (26), we obtain a set of $2n$ differential equations, which can be solved with the aid of the following.

Proposition 4.1. Let $x_i^*(t)$ and $\phi_i(t), i = 1, 2, \ldots, n$, be the solutions to (22) and (26), with q_i^* given by (28). In order for $x_i^*(t)$, $i = 1, 2, \ldots, n$, to be optimal, the following must be true.

I. $\lim_{t\to\infty} \phi_i^*(t) = 0, \qquad i = 1, 2, \ldots, n,$

if $\lim_{t\to\infty} \sum_{i=1}^{n} x_i^*(t) = \hat{X} < X_m.$

II. $\lim_{t\to\infty} \phi_i^*(t) = \alpha < 0, \qquad i = 1, 2, \ldots, n,$

if $\lim_{t\to\infty} \sum_{i=1}^{n} x_i^*(t) = X_m.$

Let X_i be such that $P_i'(0) = (\partial C_i/\partial q_i)(X_i, 0)$. From (27) and Proposition 4.2, we draw the following conclusions.

In Case I, we have $\max_i X_i = \hat{X} \leq X_m$, and each firm exploits until the total amount of production reaches X_i. There is a nonzero probability $F(\hat{X})$ that the resource will not be depleted.

In Case II, we have $\max_i X_i > X_m$. Some firms exploit until the total amount of production reaches X_i ($<X_m$) while others continue until the resource is depleted.

In either case, we can easily see that $\phi_1(t) = \phi_2(t) = \cdots = \phi_n(t)$. Moreover, for $X \in S = \bigcap_{i=1} S_i$, we have

$$\frac{\gamma_i U_i'(\Pi_i^*)}{\gamma_j U_j'(\Pi_j^*)} = \frac{\partial \Pi_i^*/\partial q_i}{\partial \Pi_j^*/\partial q_j} = \frac{P_i'(q_i^*) - (\partial C_i/\partial q_i)(X, q_i^*)}{P_j'(q_j^*) - (\partial C_j/\partial q_j)(X, q_j^*)}, \qquad i, j = 1, 2, \ldots, n. \tag{29}$$

A typical utility function is $U_i = \Pi_i^\alpha, i = 1, 2, \ldots, n, 0 < \alpha < 1$. With such functions, (29) reduces to

$$\left(\frac{\Pi_i^*}{\Pi_j^*}\right)^{1-\alpha} = \frac{\gamma_i(\partial \Pi_i^*/\partial q_i)}{\gamma_j(\partial \Pi_j^*/\partial q_j)}. \tag{30}$$

Equation (30) can be considered an equilibrium condition when all the q_i^* are positive. At such an equilibrium, the ratio of the revenues of any two firms is proportional to $(\gamma_i/\gamma_j)^{1/(1-\alpha)}$ and also proportional to the ratio of the marginal revenues (with respect to exploitation rates) raised to the powers $1/(1-\alpha)$. Suppose that $\Pi_i = \Pi_j = \Pi$; then (30) becomes

$$\frac{\partial\Pi^*/\partial q_i}{\partial\Pi^*/\partial q_j} = \frac{\gamma_j}{\gamma_i}, \qquad i, j = 1, 2, \ldots, n, \tag{31}$$

from which the ratio q_i^*/q_j^* can be determined. Since Π is monotonic increasing and concave in q when $X \in S$, we can easily see that $q_i^* > q_j^*$ when $\gamma_i > \gamma_j$.

Consider a two-firm case where $\alpha = \frac{1}{2}$, $P_1 = 2q_1$, $P_2 = q_2$, $\gamma_1 = \gamma_2$, $C_i = Xq_i$, $i = 1, 2$, and $X_m = 1$. Then (30) reduced to

$$\frac{dx_1}{dx_2} = \frac{2 - X}{1 - X} = \frac{2 - x_1 - x_2}{1 - x_1 - x_2}. \tag{32}$$

The solution of this differential equation is

$$3 - 2(x_1 + x_2) = Ke^{2(x_2 - x_1)}, \tag{33}$$

which represents a family of curves inside the right triangle $\{\binom{x_1}{x_2} \mid x_1 \geq 0, x_2 \geq 0, \text{ and } x_1 + x_2 \leq 1\}$ and terminating at the hypotenuse.

ACKNOWLEDGMENT

This research was partially supported by the National Science Foundation under Grant number SOC76-19580.

Appendix

Proposition 4.1 can be proved by first considering a finite horizon optimization problem; i.e., instead of (23), we have

$$\tilde{J}_i = \int_0^T e^{-\delta t} U_i(\Pi_i) F(X)\, dt, \qquad i = 1, 2, \ldots, n, \tag{A1}$$

where t is a fixed constant. Let $\tilde{q}_i^*(t, T)$ and $\tilde{x}_i(t, T)$, $i = 1, 2, \ldots, n$, be the optimal controls and solutions respectively for such a problem. Also let $\tilde{\phi}_i(t, T)$ be the adjoint variables. The usual transversality condition then yields the following:

Case I. $\tilde{\phi}_i(T, T) = 0, \qquad i = 1, 2, \ldots, n,$

$$\text{if } \sum_{i=1}^{n} x_i^*(T, T) < X_m.$$

$$\text{Case II.}\quad \tilde{\phi}_1(T, T) = \tilde{\phi}_2(T, T) = \cdots = \tilde{\phi}_n(T, T) < 0 \quad \text{if} \quad \sum_{i=1}^{n} \tilde{x}_i^*(T, T) = X_m.$$

Next, suppose that $\tilde{x}^*(t, T)$ and $\tilde{q}^*(t, T)$ each converges uniformly in t as $T \to \infty$. Then it can be shown that

$$\lim_{T\to\infty} \tilde{x}_i^*(t, T) = x_i^*(t), \tag{A2}$$

$$\lim_{T\to\infty} \tilde{q}_i^*(t, T) = q_i^*(t), \tag{A3}$$

and

$$\lim_{t\to\infty} \lim_{T\to\infty} \tilde{\phi}_i(t, T) = \lim_{T\to\infty} \tilde{\phi}_i(T, T) = \lim_{t\to\infty} \phi_i(t). \tag{A4}$$

Upon taking limits as $T \to \infty$, Cases I and II reduce to Cases I and II in Proposition 4.2, respectively. Proof of convergences is omitted here.

Reference

1. Liu, P. T., *Optimum Exploitation of an Exhaustible Resource under Uncertainty*, Technical Report, Mathematics Department, University of Rhode Island, August 1978.

Supplementary References

Cropper, M. L., Regulating activities with catastrophic environmental effects, *Journal of Environmental Economics and Management*, Vol. 3, pp. 1–15, 1976.

Liu, P. T. and Sutinen, J. G., Pareto, optimal leasing, and investment policies for a publicly-owned exhaustible resource, *New Trends in Dynamical Systems*. Edited by A. Marzollo, Academic Press, New York, 1979.

Loury, G. C., *The Optimum Exploitation of an Unknown Reserve*, Discussion Paper No. 255, Center for Mathematical Studies in Economics and Management Science, Northwestern University, November 1976.

10

Royalties and the Extraction Rate of an Exhaustible Resource

JON G. SUTINEN

1. Introduction

An owner of an extractive resource typically does not extract the resource himself, choosing instead to sell extraction rights in the marketplace. For example, the United States government regularly sells extraction rights to coal, uranium, and petroleum on federal lands. A common feature of most extraction rights leases is payment of a royalty to the owner plus, possibly, a fixed-sum payment.†

The effects of royalties and other leasing arrangements on the intertemporal extraction path is not well understood. Several studies have evaluated the efficiency of alternative leasing arrangements [e.g., see Devanney (Ref. 1), Leland (Ref. 2), Logue, Sweeney, and Willnet (Ref. 3), but have ignored the fundamental dynamic character of the extractive resource problem. Kalter, Tyner, and Hughes (Ref. 4) construct a dynamic model of leasing an exhaustible resource but fail to account for uncertainty, a prevasive feature of extractive resource markets, and they do not consider efficiency issues. Exhaustible resource exploitation under price uncertainty has been treated by Weinstein and Zeckhauser (Ref. 5) and Lewis (Ref. 6) and stock uncertainty by Cropper (Ref. 7) and Loury (Ref. 8). However, there is no study of leasing an

† I adopt the conventional definition of a royalty payment: a sum equal to a predetermined fraction of total realized revenue in the period.

Jon G. Sutinen • Department of Resource Economics, University of Rhode Island, Kingston, Rhode Island.

exhaustible resource under uncertainty that examines intertemporal efficiency.

This paper extends the conventional model of exhaustible resource extraction by incorporating leasing under uncertainty. The intertemporal extraction paths resulting from alternative leasing arrangements are evaluated for their economic efficiency.

2. Assumptions

The model I consider contains three sets of actors: resource owners, producing firms, and consumers of the resource. All markets are competitive, no externalities exist, no inventories are held, and the private and social discount rates are equal.

There is a known, finite number of resource owners, each possessing a known quantity of the resource stock, $\bar{x}$. Therefore, there is a known quantity of the resource in the economy, $\bar{X}$.† And the quality of each deposit is assumed to be the same. All owners are assumed to have identical preferences, each risk averse with a utility function $U_1(\phi_t)$, where $U_1' > 0$, $U_1'' < 0$, and ϕ_t represents an owner's income in period t.

The producing firms are also assumed identical and risk averse, each with a utility function $U_2(\Pi_t)$, where $U_2' > 0$, $U_2'' < 0$. A firm's income in period t is given by $\Pi_t = p_t y_t - c y_t - r_t y_t$, where y_t is the amount of the resource extracted in t by the firm, p_t the market price of extracted resource, c the per unit cost of extraction, and r_t the price of the right to extract one unit of the resource (i.e., the rental price of the resource *in situ*).

Consumers' willingness to pay for a unit of the resource is assumed given by $p_t = d_t(q_t)\mu_t$, where $d_t' < 0$, q_t is the quantity demanded in t, and μ_t is a nonnegative random variable, identically and independently distributed over time with a mean value $\bar{\mu}_t = 1$.

3. Optimal Allocation

To establish a reference by which to measure the relative efficiency of alternative outcomes I assume that it is socially desirable to maximize expected discounted surplus over a finite horizon,

$$E\left\{\int_0^T e^{-\delta t} \int_0^{q_t} [d_t(\varepsilon_t)\mu_t - c]\, d\varepsilon_t\, dt\right\},$$

subject to $\int_0^T q_t\, dt = \bar{X}$, where δ is the social discount rate.

† Stock uncertainty is ignored since, as it turns out, it is irrelevant to the issues addressed in this paper.

A necessary condition for the optimal extraction sequence is

$$\bar{p}_t - c = \lambda e^{\delta t}, \qquad 0 \leq t \leq T, \tag{1}$$

where $\bar{p}_t = d_t(q_t)\bar{\mu}_t$, the expected price in period t. Differentiating (1) with respect to time gives

$$\frac{\dot{q}_t}{q_t} = \frac{\delta(\bar{p}_t - c)}{d_t'(q_t)q_t} < 0, \tag{2}$$

where $\dot{q}_t = dq_t/dt$. Equation (2) provides the basis for evaluating alternative leasing arrangements.

4. No Risk Sharing: The Fixed Rental Lease

In the first set of leasing arrangements considered, a producing firm pays a deterministic price r_t to a resource owner for the right to extract a unit of the resource. In a competitive market for extraction rights resource owners take the sequence $\{r_t\}$ as given and choose a sequence of extraction rights to sell that maximizes their discounted income

$$\int_0^T e^{-\delta t} r_t q_t \, dt,$$

subject to $\int_0^T q_t \, dt \leq \bar{X}$. The first-order condition for this problem yields the familiar condition

$$r_t = r_0 e^{\delta t}. \tag{3}$$

Clearly an owner would sell nothing for a price

$$r_t' < r_t = r_0 e^{\delta t}$$

and would sell all his rights for a price

$$r_t'' > r_t = r_0 e^{\delta t}.$$

Therefore, supply is perfectly elastic in period t at price r_t.

The derived demand for extraction rights by the producing firms is a result of each firm's maximizing its discounted expected utility of income

$$\int_0^T e^{-\delta t} E\{U_2(p_t y_t - c y_t - r_t y_t)\} \, dt.$$

A firm's willingness to pay in each period is given by the first-order condition

$$E\{U_2' \cdot (p_t - c - r_t)\} = 0,$$

or, by rearranging,

$$r_t = \frac{E\{U_2'\mu\}}{E\{U_2'\}}\bar{p}_t - c. \tag{4}$$

Market equilibrium is determined by setting the demand price (4) equal to the supply price (3):

$$\frac{\bar{p}_t}{\Delta_2} - c = r_0 e^{\delta t}, \tag{5}$$

where $\Delta_2 = E\{U_2'\}/E\{U_2'\mu\}$.

Differentiating (5) with respect to time, and recalling that $\bar{p}_t = d_t(q_t)$, yields

$$\frac{\dot{q}_t}{q_t} = \frac{\delta[\bar{p}_t - c\Delta_2]}{q_t[d_t'(q_t) - (\bar{p}_t/\Delta_2)(\partial\Delta_2/\partial q_t)]},$$

which, if $\partial\Delta_2/\partial q_t$ is assumed to be small, becomes

$$\frac{\dot{q}_t}{q_t} = \frac{\partial[\bar{p}_t - c\Delta_2]}{d_t'(q_t)q_t} < 0. \tag{6}$$

Equation (6) describes the extraction path resulting under a lease in which the producing firms bear all the risk, since a fixed rental is paid the owners.

The difference between the optimal extraction path (2), and the extraction path under a fixed rental (6) is the presence of Δ_2 in (6). Elsewhere (Ref. 9) I have shown that $\Delta_2 > 1$. Therefore, (6) is greater than (2), implying that the extraction rate under a fixed rental falls more slowly than is optimal. This result is represented graphically in Fig. 1, where the optimal path given by (2) is labeled 0 and the fixed rental path given by (6) is labeled A.

Since (6) holds everywhere and the entire stock is extracted, the production horizon is longer and the extraction rate lower in the early

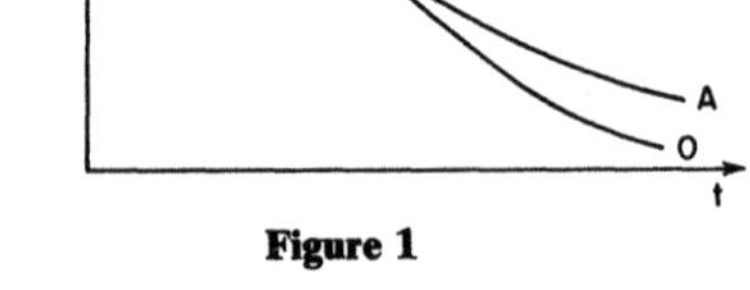

Figure 1

periods than is optimal. If firms are risk neutral, $\Delta_2 = 1$ and extraction under a fixed rental lease would be optimal. This also implies that the rate of resource depletion increases as one moves from risk-averse to risk-neutral firms, a result coinciding with that of Lewis (Ref. 6).

5. Risk-Sharing Leases

To consider risk-sharing leases I let the payment to a resource owner per unit of the resource be equal to

$$\alpha_t p_t + \beta_t,$$

where $0 \leq \alpha_t \leq 1$ and $\beta_t \gtreqless 0$.† When $\alpha_t = 0$, then $\beta_t = r_t$, the deterministic resource rental price. (Of course, α_t has the interpretation of a royalty rate.)

The lease arrangements α_t and β_t are assumed to be negotiated to the point at which neither party can be made better off without the other's being made worse off. The negotiated arrangements α_t^* and β_t^* are determined, therefore, by maximizing

$$\int_0^T e^{-\delta t} E\{U_1(\phi_t)\}\, dt$$

subject to

$$\text{(i)} \quad \int_0^T e^{-\delta t} E\{U_2(\Pi_t)\}\, dt = \int_0^T e^{-\delta t} \bar{U}_2(t)\, dt,$$

$$\text{(ii)} \quad \int_0^T q_t\, dt \leq \bar{X},$$

where $\bar{U}_2(t)$ is some level of utility at least as great as the minimal level necessary to keep the firm in the industry, $\phi_t = (\alpha_t p_t + \beta_t) y_t$, and $\Pi_t = p_t y_t - c y_t - (\alpha_t p_t + \beta_t) y_t$.

The first-order conditions determining α_t^* and β_t^* are

$$E\{U_1' p_t\} q_t - \Psi E\{U_2' p_t\} q_t = 0,$$
$$E\{U_1'\} q_t - \Psi E\{U_2'\} q_t = 0,$$

where Ψ is the multiplier on constraint (i). Eliminating Ψ results in

$$\frac{E\{U_1' p_t\}}{E\{U_2' p_t\}} = \frac{E\{U_1'\}}{E\{U_2'\}}. \tag{7}$$

† I restrict the present discussions to linear risk-sharing payment systems. Leland (Ref. 2) formulates a model with general payment systems and derives the conditions under which a linear system would be optimal. The results developed in this paper would not change if nonlinear payment systems were allowed.

Elsewhere (Refs. 9, 10) I have shown that (7) yields a share $0 \le \alpha_t^* \le 1$ for risk-averse parties. If the firm is risk neutral and the owner risk averse, then $\alpha_t^* = 0$; and if the owner is neutral and the firm risk averse, then $\alpha_t^* = 1$. Given the negotiated arrangements α_t^* and β_t^*, the firm's demand for extraction rights is given by

$$\alpha_t^* \bar{p}_t = \bar{p}_t - (c + \beta_t^*)E\{U_2'\}/E\{U_2'\mu\},$$

and the resource owners' supply of rights by

$$\alpha_t^* \bar{p}_t = \frac{\Lambda e^{\delta t} - \beta_t^* E\{U_1'\}}{E\{U_1'\mu\}},$$

where Λ is the multiplier on resource constraint faced by the owners. Combining these last two expressions and using (7) yields the market equilibrium condition

$$\bar{p}_t = \frac{\Lambda e^{\delta t} + cE\{U_1'\}}{E\{U_1'\mu\}},$$

which, when differentiated with respect to time, gives

$$\frac{\dot{q}_t}{q_t} = \frac{\delta(\bar{p}_t - c\Delta^*)}{d_t'(q_t)q_t}, \tag{8}$$

where $\Delta^* = E\{U_1'\}/E\{U_1'\mu\} = E\{U_2'\}/E\{U_2'\mu\}$ at α_t^* and β_t^*, and assuming that $\partial\Delta^*/\partial q_t$ is small. The extraction path under the share lease, as given by (8), can be compared with the path of the fixed rental lease (6) and the optimal path (2). Both (8) and (6) have a Δ-term that distinguishes them from (2).

The extraction path given by (8) is superior to the extraction path under a fixed rental (6), since $\Delta^* < \Delta_2$ and the path in (8) is nearer the optimal path in (2) than is the path in (6). The proof of this proposition is straightforward. First consider the trade-off between α_t and β_t at constant levels of expected utility for resource owners and firms, $\bar{U}_1$ and $\bar{U}_2$, respectively. The envelope theorem allows q_t to be treated as a constant in each case; therefore the marginal rate of substitution between α_t and β_t for an owner is

$$\left.\frac{d\beta_t}{d\alpha_t}\right|_{\bar{U}_1} = -\frac{E\{U_2' p_t\}}{E\{U_2'\}} = -\frac{\bar{p}_t}{\Delta_1} < 0, \tag{9}$$

where

$$\left.\frac{d^2\beta_t}{d\alpha_t^2}\right|_{\bar{U}_1} = -\frac{E\{U_1'' \cdot (p_t + d\beta_t/d\alpha_t)^2 q_t\}}{E\{U_1'\}} > 0.$$

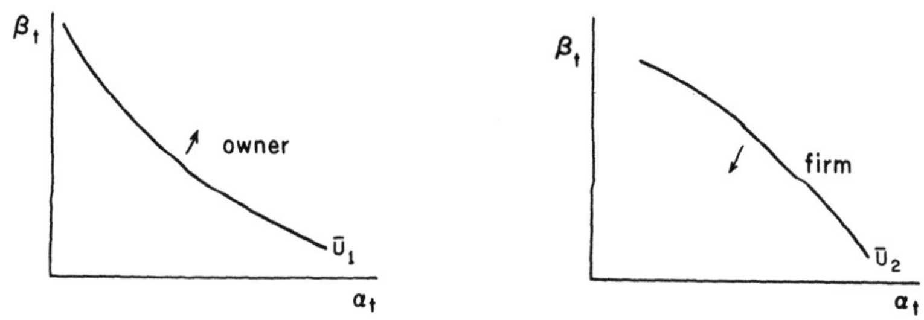

Figure 2

The marginal rate of substitution for a firm is

$$\left.\frac{d\beta_t}{d\alpha_t}\right|_{\bar{U}_2} = -\frac{E\{U_2'p_t\}}{E\{U_2'\}} = -\frac{\bar{p}_t}{\Delta_2} < 0, \tag{10}$$

where

$$\left.\frac{d^2\beta_t}{d\alpha_t^2}\right|_{\bar{U}_2} = \frac{E\{U_2''\cdot(p_t - d\beta_t/d\alpha_t)^2 q_t\}}{E\{U_2'\}} < 0.$$

The indifference curves implied by (9) and (10) are illustrated in Fig. 2, where the arrows indicate directions of higher expected utility.

Arbitrage in the marketplace yields (7), which when rearranged implies that $\Delta_1 = \Delta_2 = \Delta^*$ at α_t^* and β_t^*. Therefore, the marginal rates of substitution between the leasing arrangement α_t and β_t are equal in equilibrium, i.e.,

$$\left.\frac{d\beta_t}{d\alpha_t}\right|_{\bar{U}_1} = \left.\frac{d\beta_t}{d\alpha_t}\right|_{\bar{U}_2}.$$

This condition describes a contract curve such as the one illustrated in Fig. 3. Referring next to Fig. 4, observe that a firm's indifference curve at

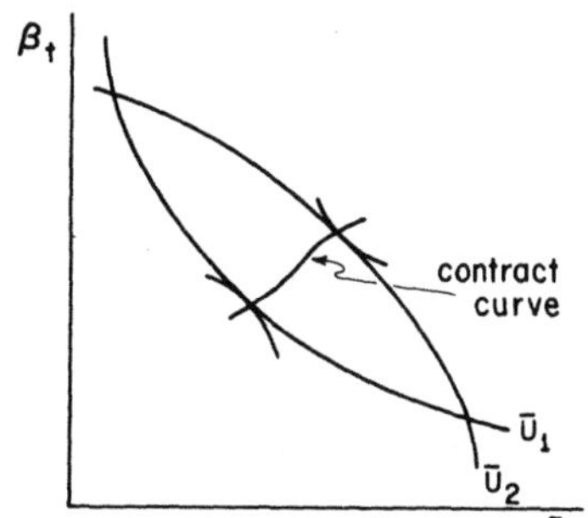

Figure 3

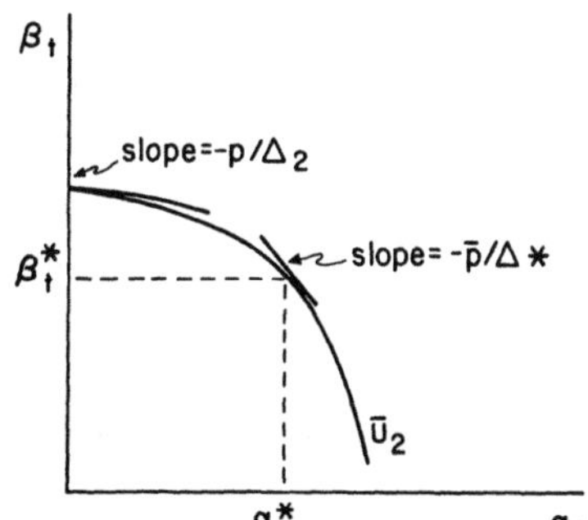

Figure 4

(α_t^*, β_t^*) has a greater absolute value than at $(0, r_t)$, the fixed rental solution. Therefore, Δ_2 for the fixed rental is greater than Δ^*, and the proposition is proved.

Further note that when both firms and owners are risk neutral, both shared and fixed rental leases yield the optimal extraction path. If only firms are risk neutral, the market will yield a fixed rental lease; if only owners are risk neutral, a fixed payment to firms will be made with the owners bearing all the risk.

6. Leasing with Stock Uncertainty

In this section we examine the implications for leasing arrangements of an uncertain resource stock. It turns out that share leasing provides no advantages with stock uncertainty. The analysis closely follows Cropper (Ref. 7).

I assume that a resource owner possesses an initial, unknown amount of the resource represented by $\hat{x}$, a continuous random variable satisfying $\mathrm{pr}\{\hat{x} > x\} = F(x)$. $F(0) = 1$, $F(x) = 0$ for $x \geq X_m$, and $F'(x) < 0$ for $0 \leq x \leq x_m$ where x_m is defined as the maximum possible initial amount of the stock. Define x_t to be the accumulated amount of extraction as of period t, i.e.,

$$x_t = \int_0^t q_s \, ds,$$

where q_t is the amount extracted in t. Price in each period is assumed known with certainty.

Expected utility at time t in this case becomes $F(x_t)U(t)$, since $F(x_t)\Delta t$ is the probability of nonexhaustion in the following infinitesimal interval Δt.

Equilibrium leasing arrangements are determined by

$$\max_{\{\alpha_t, \beta_t\}} \int_0^T F(x_t) U_1(\Pi_1) e^{-\delta t} \, dt$$

subject to

$$\text{(i)} \quad \int_0^T F(x_t) U_2(\phi_t) e^{-\delta t} \, dt = \int_0^T \bar{U}_t e^{-\delta t} \, dt,$$

$$\text{(ii)} \quad \dot{x}_t = q_t.$$

The necessary conditions determining α_t^* and β_t^* are

$$F(x_t)\{U_1' p_t q_t - \Psi U_2' p_t q_t\} = 0,$$
$$F(x_t)\{U_1' q_t - \Psi U_2' q_t\} = 0,$$

which reduce to simply $U_1' - \Psi U_2' = 0$. No unique (α_t, β_t) is implied by these conditons. There are no risk-sharing possibilities in the class of lease arrangements considered and hence no effect on the extraction path of alternative arrangements.

An uncertain stock does have implications for the path of extraction over time as shown by Cropper (Ref. 7) and extended by Liu in Chapter 9 of this volume.

7. Concluding Remarks

This paper has developed a model of competitive market for extraction rights under price uncertainty. The principal result of the analysis is that a class of risk-sharing leases (with a royalty and a fixed-sum payment) is superior to fixed rental leases with no risk sharing. In this context at least, the concern that royalties result in inefficient extraction levels is unfounded.

ACKNOWLEDGMENT

The research reported herein was supported by National Science Foundation grant number SOC76-19580. This paper is Contribution Number 1807 of the Rhode Island Agricultural Experiment Station.

References

1. DEVANNEY, J. W., *The OCS Petroleum Pie*, MIT Sea Grant Program, Cambridge, Mass., 1975.
2. LELAND, H., *Optimal Risk Sharing and the Leasing of Natural Resources, with Application to Oil and Gas Leasing on the OCS*, Working Paper No. 38, Graduate School of Business Administration, University of California, Berkeley, 1977.
3. LOGUE, D. E., SWEENEY, R. J., and WILLNET, T. D., Optimal leasing policy for the development of outer continental shelf hydrocarbon resources, *Land Economics*, Vol. LI, No. 3, pp. 191–207, 1975.
4. KALTER, R. J., TYNER, W. E., and HUGHES, D. W., *Alternative Energy Leasing Strategies and Schedules for the Outer Continental Shelf*, Department of Agricultural Economics, Cornell University, December 1975.
5. WEINSTEIN, M. C. and ZECKHAUSER, R. J., The optimal consumption of depletable natural resources, *Quarterly Journal of Economics*, Vol. 89, pp. 371–392, 1975.
6. LEWIS, T. R., Attitudes towards risk and the optimal exploitation of an exhaustible resource, *Journal of Environmental Economics and Management*, Vol. 4, pp. 111–119, 1977.
7. CROPPER, M. L., Regulating activities with catastrophic environmental effects, *Journal of Environmental Economics and Management*, Vol. 3, pp. 1–15, 1976.

8. Loury, G., *The Optimum Exploitation of an Unknown Resource*, mimeographed, Center for Mathematical Studies in Economics and Management Science, Northwestern University, 1976.
9. Sutinen, J. G., The rational choice of share leasing and implications for efficiency, *American Journal of Agricultural Economics*, Vol. 57, No. 4, pp. 613–621, November 1975.
10. Sutinen, J. G., *Towards a General Theory of Share Contracting*, mimeographed, Department of Resource Economics, University of Rhode Island, 1974.

Part III

Large-Scale Systems and Decentralized Control

A new trend in the development of dynamic optimization has been the study of large-scale systems. This trend has been motivated by problems in large-scale power networks and socioeconomic systems. It has received considerable attention because of its theoretical interest and practical importance.

There is no precise, well-established definition of a large-scale system (it is hardly possible to make up one either). The scope of the topic is expanding rapidly and interacting with more and more problems in other areas. Roughly speaking, a large-scale system is a collection of a number of subsystems, which are somewhat interconnected, and each having its own controller.

A team decision problem usually designates a special type of large-scale system with a single objective function to be minimized (or maximized) jointly by all controllers. Some communication among controllers is not only realistic but also essential for successful operation of such a system. The degree of communication affects each controller's behavior to a large extent. The study of information structure concerns the specification of how much each controller knows about the system and how much of this he can utilize in his decision making. The problem is basically stochastic in nature, because information measurement is almost always associated with probability distribution functions, although for some special deterministic problems, decentralization can still be achieved through decomposition in vector spaces.

In large-scale systems, interaction among subsystems is, in general, relatively weak. This weak coupling often leads to the construction of a suboptimal control law. However, sometimes subsystems are not weakly coupled, but they are similar to each other, as in the case when subsystems are identical in dynamic structures, except for differences in system parameters. The control of each subsystem can then be based on information pertaining to its own subsystem plus some aggregated signals. This is the model adopted in Aoki's paper, in which several examples in macroeconomics are explained.

In a competitive economy, decisions are also made through decentralization among numerous firms (agents). However, the degree of complexity is increased by the fact that each firm may have its own objective function, and these objective functions may be mutually conflicting. Rational decision making then rests on the definition of an optimal or equilibrium solution. Multicriteria decentralized optimization is an area that remains very much unexplored.

Another feature of a large-scale system is its hierarchial decision-making aspect. In a socioeconomic or power supply system, optimization is frequently to be carried out at several levels. For a two-level system, for example, there are a central controller and several other local controllers. In such a situation, the Stackelberg solution is often used as an approach to rational decision making. In this approach, each local controller determines an optimum control u_i^* $(i = 1, 2, \ldots, n)$ based on a central control v. The central controller (or the coordinator) is supposed to know all the objective functions and therefore anticipate all optimizing controls u_i^* $(i = 1, 2, \ldots, n)$ for each v. He then chooses v to maximize his own objective function. This is the so-called Stackelberg optimization in hierarchical decision making.

In Basar's paper, the Stackelberg solution for a class of two-person quadratic decision problems is discussed. The hierarchical solution concept allows one of the controllers (the leader) to have access to the observation and the strategy of the other decision maker (the follower). It is verified that such a decision problem admits unique equilibria under static and dynamic information structures; and in the case of dynamic information, the equilibrium strategy of the leader has the property that he can enforce a Pareto-optimal solution in his favor. This paper also includes two examples of a stochastic production model and a stochastic price model in duopoly markets when Firm 1 is taken as the leader in the decision process.

Bureaucratic systems are the most common examples of large-scale systems with hierarchical structures in modern society. In such a system, there are a central authority and several bureaus, i.e., offices or branches. In addition, there is a review authority.

Brito and Intriligator consider a game problem between two bureaus competing for resource allocation. The central authority makes decisions in the allocation, subject to possible revision by the review authority. Each agent (bureau, central authority, or review authority) is assumed to maximize its own welfare functional under a budget constraint. Aside from the hierarchical decision aspect of the problem, this problem is one of a two-person, zero-sum game in which a Nash bargaining solution is to be sought with a "threat" from the review authority.

National policies on macroeconomic systems in this country are dictated by decentralized control processes. An obvious example is the separation of monetary and fiscal policies. Different authorities manipulate different controls and frequently have conflicting target sets. In Myoken's paper, decentralized policies for different target levels are investigated. The difference between the equilibrium value and the target value is called the bias. It is shown that such biases for static and dynamic models are identical. Of particular interest is the illustrative example of Klein's model 1, which gives a simplified description of the U.S. economy for the period 1921–1941.

As is well known, control of stochastic dynamical systems with partial observation is rather difficult. The mathematical techniques involved include estimation, identification, and adaptive control. All these require a great deal of analytical and computational manipulation. This is even more so in decentralized controls where each controller has his own observation.

As explained in the paper by Varaiya and Walrand, the concept of a state for a system hardly exists in decentralized controls because aggregation of information cannot be summarized as a sufficient statistic. Admissible decentralized control laws are defined as measurable functions with respect to several sigma fields. Hence, any optimality criterion, e.g., via dynamic programming, has to be expressed in the abstract setting of conditional expectation with respect to one single sigma algebra.

Varaiya and Walrand introduce a control law which is called locally-in-time and by-person optimal. Such a control is optimal in a narrow sense, but it is attractive since it can be characterized by a necessary condition in the form of a maximum principle. In a way, this is an optimality condition for control of a non-Markovian system.

11

Dynamics and Control of a System Composed of a Large Number of Similar Subsystems

MASANAO AOKI

1. Introduction

There are dynamic systems that are composed of many interacting subsystems and are too large or too complicated in some senses to lend themselves conveniently to the various analytical or computational procedures for control and/or estimation that we find in control and system theory literature, since these techniques require, on the whole, that states and control actions by each and every decision maker of subsystems be known to each other or to some coordinating agents.

Such "large" systems can be treated from a number of viewpoints. We can focus on the information or decision problem facing the central or coordinating agents charged with controlling "aggregate" behavior of such systems in some sense. Alternatively we can focus on individual subsystems and their information and decision problems.†

† Centralized or decentralized planning in economic literature is relevant to these viewpoints. Large systems may have central or coordinating agents who are charged with controlling aggregate or average dynamic behavior of the total system when subsystem characteristics are not known precisely. A central decision maker needs to receive information on the subsystem states and/or characteristics, or to estimate the subsystem states, from the macro (aggregate) system observations which are in known deterministic or stochastic relation to the subsystems. Several examples are mentioned, and some preliminary analysis for such estimation problems is carried out in Aoki (Ref. 1). Difficulties arise when central agents cannot collect the necessary information on the

MASANAO AOKI • Department of System Science, University of California, Los Angeles, California.

With another view, the problem is formulated as stochastic (contact) processes of interacting particles or agents treating subsystem decision functions as known functions of subsystem and/or total system states. (See Liggett, Ref. 2, for example.) Regardless of the approach that one takes, lack of information on other subsystem states and their control actions and/or information about interactions among subsystems causes severe problems in synthesizing optimal or near optimal control or estimation schemes for subsystem controllers in larger systems. This is because "effectiveness" or "goodness" of decisions taken by a control agent (decision maker) is often determined as a function of all other decisions taken (more or less) simultaneously by all other agents in the system, since their decisions jointly determine the outcome and/or the next state of the aggregate system.

As Aoki and Li (Ref. 3) have shown, it is an exception rather than the rule that decision makers of subsystems can satisfactorily estimate the other subsystem states and/or controls from their own observation records alone. Thus, decisions (controls) taken by other decision makers need be known for "optimal" decisions by any agent. An individual agent cannot in general infer correctly decisions taken by other agents from his own information set unless he is provided with extensive communication links to other decision makers.‡ (This fact remains true whether decision makers have identical or similar goals as in team problems or more generally conflicting goals.)

We do not yet have a complete theory of control or estimation of large systems. We consider in this paper a special class of large systems that are composed of many "weakly" interacting subsystems whose decision functions are parametrically specified. For example, subsystems may be identical in structure, each differing from others only in having different system parameters or information sets due to different realizations of some stochastic processes to which they are subjected. We consider a subclass of such large systems in which interaction among subsystems is not overly strong§ and comes from the fact that in each subsystem decisions are made in response to information that contains, in addition

---

subsystems and must decide without precise information on the states and control actions of subsystems.

In decentralized systems, i.e., systems without central agents, similar informational difficulties confront subsystem decision units.

§ These communication links are not likely to be found in existing systems and/or are much too expensive to install for new large systems (some technological systems may be exceptions to this general statement). In many large systems, then, we can safely assume that decision makers do not have adequate information to estimate other subsystem states and/or controls with great accuracy.

‡ In contact processes individuals' interactions with "neighbors" play a dominant role. Our focus is rather on interaction of individuals with the "average" or macro signals.

to information pertaining to his own subsystem, some common aggregate signals. Thus, subsystems are influenced to some degree by the common set of aggregate signals, and they in turn jointly generate the new aggregate signals.

Thus, we shift our focus from exact description of the total system's dynamic behavior and consider a collection of many similar interacting subsystems in which individual subsystems are sufficiently small in their interaction with other systems that only some "average" or aggregate effects of these subsystem actions are felt by individual subsystems, and we describe them together by characterizing the resulting distribution of subsystem traits. More specifically, we examine the first two moments of the equilibrium distribution of subsystem states and their response to distribution of parameters characterizing subsystem dynamics and to macro signals.†

The notion that some aggregate signals affect decision processes of individual subsystems is not new. Many economic phenomena may be modeled in this manner. For example, pricing and output decisions by firms in some industry may be influenced by past histories of observed market prices and outputs sold, which every firm in the industry observes and to which every firm similarly reacts. This example is analyzed in Section 2. We are interested in the "average" behavior of subsystems over time and in the spread of possible subsystem behavior consistent with the macro signals, among others. The viewpoint of this paper is therefore related to those of Aoki (Ref. 1) and Fukao (Ref. 7). As another example, observe that it is postulated in many macroeconomic models that aggregate excess demand determines the rate of change of a price level. These models, however, do not usually consider distribution of economic agents, assuming them to be identical to a representative agent. In many cases, this simplification is justifiable. However, we are interested in avoiding this simplification, and we treat some of the distributional effects. A growing body of economic literature on distribution focuses primarily on description of distributions and does not share the same focus as this paper on control and intertemporal dynamics.

† There is similarity with the economic literature on distribution of wealth and income. See, for example, J. E. Stiglitz (Ref. 4), F. L. Pryor (Ref. 5), or J. Conlisk (Ref. 6). Our approach is similar in aim to Stiglitz, and our analysis may be regarded as providing some analytical insight into simulation results. Stiglitz isolated the different economic forces that tend to make the distribution of wealth egalitarian in the long run and those that tend to make wealth unevenly distributed by examining the rates at which the wealth (and income) of different groups grows. He examined, for example, a heterogeneous labor force as one of the forces for uneven wealth distribution. (He did not actually characterize the resulting distribution except for the case of efficiency of labor.) He assumed the propensity to save to be the same for all the groups. We treat in one example the propensity to save as one of the major sources for uneven distribution and characterize the first and second moments of the resulting distribution.

The paper begins by examining several examples in some detail in order to motivate a general class of systems formulated later in the paper.

2. Suggestive Examples

Before giving a general problem formulation, I discuss several examples that will clarify and motivate the general approach to this class of systems to be given later. To capture the idea that every subsystem is subjected to macro signals that depend on the aggregates of subsystem states and decisions made earlier in a simple way, we adopt a discrete-time formulation.

Example 2.1. To illustrate interactions of decisions by subsystems, we use the example of Aoki (Ref. 8). There are n firms in an industry producing a homogeneous perishable good. (This assumption of perishability is made to avoid the complication of inventory. Although this simplification may not be very realistic, the example serves its purpose of suggesting a general class of dynamic systems to be investigated later.)

Here the manager of a firm must decide how much to change the current output. The output of firm i at time t, q_t^i, is changed by Δq_t^i so that on day $t + 1$, firm i produces

$$q_{t+1}^i = q_t^i + \Delta q_t^i.$$

The difficulty arises because the price P_{t+1}, at which q_{t+1}^i is sold, is not known. It is a function of the decisions of all the firms in this particular industry plus possible exogeneous random disturbance. The price is assumed to be governed by

$$P_{t+1} = \alpha - \delta Q_{t+1} = P_t - \delta\eta_t^i - \delta\Delta q_t^i,$$

where

$$\eta_t^i = \sum_{j\neq i} \Delta q_t^j$$

represents effects of other firms' decisions on the price. We assume that P_t (or its noisy observation) becomes known to every firm at the end of the tth period. There are several interesting learning or information problems in connection with estimating other firms' behavior. Since they are discussed elsewhere (Aoki, Ref. 9) and are not central to this investigation, I shall not go into this aspect. If we assume that the decisions of the other firms, i.e., η_t^i, are known, then myopic or one-period profit optimization leads firm i to change its output by

$$\Delta q_t^i = [P_t - \sigma_i(q_t^i) - \delta\eta_t^i]/(b_i + 2\delta), \tag{1}$$

where, assuming a linear marginal cost function,

$$\sigma_i(q) = a_i + (b_i + \delta)q$$

is firm i's supply price schedule. The parameters a_i and b_i determine uniquely the marginal cost curve of firm i. We assume that firm i knows a_i and b_i, but a_j and b_j, $j \neq 1$, are not known. Their statistical distribution is assumed to be known.

The state of the industry is represented by a vector $\mathbf{q}_t = (q_t^i, \ldots, q_t^n)'$. It evolves with time as

$$\mathbf{q}_{t+1} = \mathbf{q}_t + \Delta\mathbf{q}_t,$$

where

$$\Delta\mathbf{q}_t = (\Delta q_t^1, \ldots, \Delta q_t^n)'$$

are functions of a macro signal P_t or past history of P's. We assume that α and δ are known to all the firms so that Q_t is known to all the firms. We regard (1) as the decision rule for firm i when η_t^i is replaced with its estimate in whatever manner the estimate is formed.

Let $\gamma_i = (b_i + 2\delta)^{-1}$. Then the output vector $\mathbf{q}_t$ composed of the individual firm's output q_t^i, $i = 1, \ldots, n$, evolves dynamically as

$$\mathbf{q}_{t+1} = \delta(\Gamma - \boldsymbol{\gamma}\mathbf{e}')\mathbf{q}_t + \mathbf{u} - \delta\mathbf{d}_t,$$

where

$$\Gamma = \operatorname{diag}(\gamma_1 \cdots \gamma_n), \qquad \boldsymbol{\gamma} = \Gamma\mathbf{e}, \qquad \mathbf{u} = (u_1 \cdots u_n),$$

where

$$u_i = (\alpha - a_i)\gamma_i$$

and where $\mathbf{d}_t$ is a vector of decisions with its ith component equal to γ_i times firm i's estimate of η_t^i. The variable d_t^i is thus regarded as firm i's decision variable, since it determines the actual Δq_t^i under the decision rule (1). Note that a dyad $\boldsymbol{\gamma}\mathbf{e}'$ provides an interaction term. This is a feature common to all systems examined in this paper. We are interested in the influences of the individual decisions and their average on the mean output and spread of individual firms' outputs from the mean as well as in distributional effects of some exogenous signals such as α or δ, since they may be modified to some extent by the outside agencies directly or by changes in regulations relating to this particular industry. Thus, we examine, for example, whether a change in α impinges differentially on the mean and deviations of the outputs or some other measures of spreads of the outputs such as the coefficient of variation.

Although we can treat some dynamic effects (see Aoki, Ref. 9), we confine the examination of the distributional effects to the steady state in this paper.

We next derive the dynamic equation for the mean and the deviational vector of firms' outputs. Let $\bar{\gamma}$ be the mean of the γ's so that $\gamma^i = \bar{\gamma} + \zeta_i$ where $\sum_i \zeta_i = 0$ and $a_i = a + \varepsilon_i$ where $\sum_i \varepsilon_i = 0$. Then the matrix Γ can be written as

$$\Gamma = \bar{\gamma}I + \Xi, \qquad \text{where} \quad \Xi = \operatorname{diag}(\zeta_1 \cdots \zeta_n),$$

and the vector $\mathbf{u}$ becomes

$$\mathbf{u} = (\alpha - a)\bar{\gamma}\mathbf{e} - \bar{\gamma}\boldsymbol{\varepsilon} + (\alpha - a)\boldsymbol{\zeta} - \boldsymbol{\theta},$$

where

$$\boldsymbol{\varepsilon} = (\varepsilon_1 \cdots \varepsilon_n)', \qquad \boldsymbol{\theta} = (\varepsilon_1\zeta_1 \cdots \varepsilon_n\zeta_n)'.$$

The mean output m_t is defined by $\mathbf{e}'\mathbf{q}_t/n$. It is governed by

$$m_{t+1} = -(n-1)\,\delta\bar{\gamma}m_t + \delta\boldsymbol{\zeta}'\mathbf{s}_t/n + (\alpha - a)\bar{\gamma} - \mathbf{e}'\boldsymbol{\theta}/n - \delta\bar{d}_t, \tag{2}$$

where $\mathbf{s}_t = \mathbf{q}_t - m_t\mathbf{e}$, and where $\bar{d}_t = \mathbf{e}'\mathbf{d}_t/n$ is the average of the decisions across the firms. Assume the ε's and ζ's are uncorrelated (this is likely to be the case since there is no reason to believe that the variation in the intercept points of the marginal cost curves are correlated with the variations related to the slope). We assume that $\mathbf{e}'\boldsymbol{\theta}/n$ is nearly zero and drop it from (2). The deviation from the mean evolves with time as follows:

$$\mathbf{s}_{t+1} = \delta(\bar{\gamma}I + \Xi)\mathbf{s}_t - \delta(n-1)m_t\boldsymbol{\zeta} - \delta\mathbf{e}\boldsymbol{\zeta}'\mathbf{s}_t/n + \boldsymbol{\mu} - \delta\mathbf{v}_t, \tag{3}$$

where $\boldsymbol{\mu} = -\bar{\gamma}\mathbf{e} + (\alpha - a)\boldsymbol{\zeta} - \boldsymbol{\theta}$ is a mean zero constant, $\mathbf{v}_t = \mathbf{d}_t - \bar{d}_t\mathbf{e}$ is the deviation of the individual decisions from their average. Equations (2) and (3) combine to describe the dynamics of the outputs

$$\begin{bmatrix} m_{t+1} \\ \mathbf{s}_{t+1} \end{bmatrix} = \Phi\begin{bmatrix} m_t \\ \mathbf{s}_t \end{bmatrix} - \delta\begin{pmatrix} \bar{d}_t \\ \mathbf{v}_t \end{pmatrix} + (\alpha - a)\begin{bmatrix} \bar{\gamma} \\ \boldsymbol{\zeta} \end{bmatrix} - \begin{bmatrix} 0 \\ \gamma\boldsymbol{\varepsilon} + \boldsymbol{\theta} \end{bmatrix}, \tag{4}$$

where

$$\Phi = \Phi^0 + \frac{\delta}{n}\begin{bmatrix} 1 \\ -\mathbf{e} \end{bmatrix}[0, \boldsymbol{\zeta}']$$

and where

$$\Phi^0 = \begin{bmatrix} -(n-1)\,\delta\bar{\gamma} & 0 \\ -\delta(n-1)\boldsymbol{\zeta} & \delta(\bar{\gamma}I + \Xi) \end{bmatrix}.$$

Under the usual assumption of a representative firm we have $\boldsymbol{\zeta} = 0$, and Φ reduces to a block diagonal matrix without interaction between the mean variable and the deviation variable. If we assume that $\boldsymbol{\zeta} = 0$ but $\mathbf{u} \neq 0$, i.e., that the slope of the marginal cost curve is identical for all firms but the estimates of the rest of the industry's growth vary by firms, then (4) simplifies to

$$\begin{pmatrix} m_{t+1} \\ \mathbf{s}_{t+1} \end{pmatrix} = \begin{bmatrix} -(n-1)\,\delta\bar{\gamma} & 0 \\ 0 & \delta\bar{\gamma}I \end{bmatrix} \begin{pmatrix} m_t \\ \mathbf{s}_t \end{pmatrix} - \delta \begin{pmatrix} \bar{d}_t \\ \mathbf{v}_t \end{pmatrix} + (\alpha - a)\begin{pmatrix} \bar{\gamma} \\ 0 \end{pmatrix} - \bar{\gamma}\begin{pmatrix} 0 \\ \boldsymbol{\varepsilon} \end{pmatrix},$$

showing $\mathbf{v}_t$ and the variation in the intercepts points of the marginal cost curves as the sole sources of variations in the firms' outputs, and there is no interaction between m_t and $\mathbf{s}_t$. The steady-state solution exists if $(n - 1)\,\delta\bar{\gamma} < 1$ and is given by

$$\begin{pmatrix} m_\infty \\ \mathbf{s}_\infty \end{pmatrix} = (I - \Phi)^{-1}\left\{-\delta\begin{pmatrix} \bar{d} \\ \mathbf{v} \end{pmatrix} + (\alpha - a)\begin{pmatrix} \bar{\gamma} \\ 0 \end{pmatrix} - \bar{\gamma}\begin{pmatrix} 0 \\ \boldsymbol{\varepsilon} \end{pmatrix}\right\}$$

$$= \begin{bmatrix} [-\delta\bar{d} + (\alpha - a)\bar{\gamma}]/[1 + (n-1)\,\delta\bar{\gamma}] \\ -(\delta\mathbf{v} + \bar{\gamma}\boldsymbol{\varepsilon})/(1 - \delta\bar{\gamma}) \end{bmatrix}.$$

Of course $\bar{d}$ is zero if firms adjust their estimates of the growth rate of the rest of the industry. The point to note is that neither $\mathbf{v}$ nor $\boldsymbol{\varepsilon}$ appear in m_∞. The independent determination of the mean and variance disappears as we drop the assumption of representative firms. Then $\boldsymbol{\zeta}$ becomes nonzero.

We next obtain the steady-state solution to exhibit this interdependence for the general case. The actual calculation is carried out in the appendix.

In the general case, assuming $\bar{d} = 0$, the expression for the mean steady-state output is

$$m_\infty = \frac{(\alpha - a)\bar{\gamma} + \delta\tau}{1 + (n-1)\,\delta\bar{\gamma}}, \tag{5}$$

where

$$U = \boldsymbol{\zeta}'(I - \delta\Gamma)^{-1}\mathbf{v},$$

$$V = \boldsymbol{\zeta}'(I - \delta\Gamma)^{-1}\boldsymbol{\zeta},$$

$$\tau = (\alpha - a)[1 + (n-1)\,\delta\bar{\gamma}]^{-1}V - \delta U - \boldsymbol{\zeta}'(I - \delta\Gamma)^{-1}(\gamma\boldsymbol{\varepsilon} + \boldsymbol{\theta}).$$

If $\mathbf{v}$ is uncorrelated with $\boldsymbol{\zeta}$ (i.e., unless firms with similar marginal cost slopes tend to estimate the growth rate of the rest of the industry similarly), U will be zero or small in magnitude. The steady-state deviation is given by

$$\mathbf{s}_\infty/m_\infty = -\delta(n-1)(I-\delta\Gamma)^{-1}\zeta \\ -m_\infty^{-1}\{\delta(I-\delta\Gamma)^{-1}\mathbf{v} + (I-\delta\Gamma)^{-1}(\bar{\gamma}\boldsymbol{\varepsilon}+\boldsymbol{\theta}) \\ +(\alpha-a)(I-\delta\Gamma)^{-1}\boldsymbol{\zeta} + \tau\delta[n(1-\Delta)]^{-1}(I-\delta\Gamma)^{-1}\mathbf{e}\}, \quad (6)$$

where

$$\Delta = -\frac{\delta}{n}\left\{\frac{\delta(n-1)}{1+\delta(n-1)\bar{\gamma}}V + E\right\},$$

$$E = \boldsymbol{\zeta}'(I-\delta\Gamma)^{-1}\mathbf{e},$$

$$\kappa = \frac{(\alpha-a)\,\delta[n(1-\Delta)]^{-1}V}{(\alpha-a)\bar{\gamma} - \delta[n(1-\Delta)]^{-1}U}.$$

Thus, m_∞ deviates from the representative firm case by a quantity proportional to U, a generalized variance of $\boldsymbol{\zeta}$'s.

Since $(\alpha - a)/m_\infty$ is approximately equal to $[1 + (n-1)\,\delta\bar{\gamma}]/\bar{\gamma}$ if U is small, (6) shows that $\mathbf{s}_\infty/m_\infty$ has a component roughly proportional to $\boldsymbol{\zeta}$ with the positive coefficient if $1/\bar{\gamma} > \delta(n-1)\kappa$ and with the negative coefficient if $1/\bar{\gamma} < \delta(n-1)\kappa$. When U is small in magnitude, the former is likely. It implies that the component of $\mathbf{s}_\infty$ is such that firms with high ζ's tend to deviate positively from m_∞. The other component of $\mathbf{s}_\infty$ is roughly proportional to $\delta\mathbf{v} + \bar{\boldsymbol{\gamma}}\boldsymbol{\varepsilon} + \boldsymbol{\theta} + \tau\delta\cdot[n(1-\Delta)]^{-1}\mathbf{e}$, combining effects of $\mathbf{v}$ (disparate estimates held by firms), $\boldsymbol{\varepsilon}$ and $\boldsymbol{\theta}$ (different parameters in the marginal cost curves), and V (generalized variance of γ's). Deviations in any of these variables from their means produce negative, i.e., opposite, deviations in $\mathbf{s}_\infty$. One useful measure of the deviation is the "coefficient of variation" $(\mathbf{s}_\infty s_\infty)^{1/2}m_\infty$. We calculate its square as $\mathbf{s}_\infty\mathbf{s}_\infty/m_\infty^2 \approx \delta^2(n-1)^2\zeta'(I-\delta\Gamma)^{-2}\boldsymbol{\zeta} + O(m_\infty^{-1})$, if we assume that the correlations among $\mathbf{v}$, $\boldsymbol{\varepsilon}$, $\boldsymbol{\theta}$ in $\mathbf{e}$ are small enough compared with m_∞^{-1}. The expression becomes exact as m_∞ becomes large (for example, by letting $\alpha - a$ become large). The coefficient of variation is thus seen to be independent of $\mathbf{v}$ provided $\mathbf{v}'(I-\delta\Gamma)^{-2}\mathbf{v}/m_\infty \to 0$ as $m_\infty \to 0$ and similarly for $\boldsymbol{\varepsilon}$, $\boldsymbol{\theta}$, and $\mathbf{e}$.

Example 2.2. This example is based on Conlisk (Ref. 6) and presents steady-state analysis of his simplified model (ignoring bequest, for example). Such analysis seems useful, since most recent work on micro simulation lacks analysis of distributional effects of microeconomic policies. See also footnote†, p. 185.

Consider a society with M households. We adopt a drastic simplification and assume that the ith household at time period t (tth generation) is characterized by pretax income Y_t^i, tax T_t^i, nonhuman capital assets K_t^i, and hours of employment N_t^i. We use the same letters without superscripts to refer to the average of the same variable over the households. For example,

$$Y_t = \sum_i Y_t^i/M$$

is the average pretax income, and so on.

These subsystem variables are related to each other by

$$Y_t^i = r_t K_t^i + w_t N_t^i, \tag{7}$$

$$T_t^i = \tau(Y_t^i - \beta Y_t), \tag{8}$$

$$K_{t+1}^i = (1 - \delta)K_t^i + s^i(Y_t^i - T_t^i), \tag{9}$$

$$N_t^i = \gamma_1 N_{t-1} + \gamma_2 N_{t-1}^i + \delta_1 Y_{t-1} + \delta_2 Y_{t-1}^i. \tag{10}$$

We ignore errors in these equations for simplicity. (Generally, there are errors in the variables as well as equation errors.) Equation (7) shows that the income of the ith household consists of the return on the capital stock it owns, where the rate of return r_t is common to all households, and of the labor income, where w_t is the common wage rate. Equation (8) means that the tax it pays is computed based on the exemption related to the average income. τ and β are the macroeconomic instruments of this system. Equation (9) shows the manner in which a household accumulates assets, while (10) gives the dynamics of employment. The saving rate s^i is the decision variable of the ith household.

The main point ot note is that the equations for subsystems variables contain some aggregate variables common to all households. The relations among aggregate variables are also known:

$$Y_t = f(K_t/N_t)N_t, \tag{11}$$

$$r_t = f'(K_t/N_t), \tag{12}$$

$$w_t = (Y_t - r_t K_t)/N_t, \tag{13}$$

where (11) is a convex macro production function which is assumed to exist. Equations (12) and (13) relate the rate of return to owning capital stock to marginal productivity, and similarly for the wage rate. Individual households perform intertemporal optimization, given these macro variables, i.e., they are treated as price takers, since their individual decisions do not influence macro variables in any appreciable degree.

We are interested in characterizing the distribution of wealth of households in terms of the distributions of the s^i and in examining the ways that distribution changes with respect to the macro decision variables τ and β.

By aggregating (7)–(9), we obtain the dynamic equation for the $k_t = K_t/N_t$, where $K_t = \sum_i K_t^i/M$. Noting that we have $Y_t/N_t = f(k_t)$ from

(11) and that

$$N_{t+1}/N_t = \gamma + \delta f(k_t), \tag{14}$$

where $\gamma = \gamma_1 + \gamma_2$,

$$k_{t+1} = \frac{(1-\delta)k_t + \bar{s}[\tau\beta + (1-\tau)]f(k_t) + \xi_t}{\gamma + \delta f(k_t)}, \tag{15}$$

where

$$\xi_t = (1-\tau)[r_t \operatorname{cov}_t^1 + w_t \operatorname{cov}_t^2],$$

$$\bar{s} = \sum \frac{s^i}{M}, \qquad \operatorname{cov}_t^1 = \sum_i \frac{(s^i - \bar{s})\kappa_t^i}{M}, \qquad \operatorname{cov}_t^2 = \sum_i \frac{(s^i - \bar{s})\eta_t^i}{M},$$

and where we define the deviational variables

$$\kappa_t^i = k_t^i - k_t, \qquad \text{where} \quad k_t^i = K_t^i/N_t,$$

and

$$\eta_t^i = N_t^i/N_t - 1.$$

Here τ may be taken as the macro control variable and it is reflected in the value of ξ_t. The deviations of the individual household from the average are governed by a coupled nonlinear equation:

$$\begin{pmatrix} k_{t+1} \\ \eta_{t+1} \end{pmatrix} = \Phi_t \begin{pmatrix} k_t \\ \eta_t \end{pmatrix} + \phi_t \begin{pmatrix} \mathbf{s} - \bar{s}\mathbf{e} \\ 0 \end{pmatrix}, \tag{16}$$

where Φ_t is a $(2M \times 2M)$ matrix with submatrices†

$$\begin{aligned}
\Phi_{t,11} &= [(1-\delta)I + (1-\tau)r_t S]/a_t - (1-\tau)r_t \mathbf{ee}'S/a_t M, \\
\Phi_{t,12} &= (1-\tau)w_t S/a_t - (1-\tau)w_t \mathbf{ee}'S/a_t M, \\
\Phi_{t,21} &= \delta_2 r_t I/a_t, \\
\Phi_{t,22} &= (\gamma_2 + w_t \delta_2)I/a_t, \\
a_t &= \gamma + \delta f(k_t), \qquad S = \operatorname{diag}(s^1, \ldots, s^n), \\
\Phi_t &= [1 - \tau(1-\beta)]f(k_t)/a_t, \\
\mathbf{s} &= S\mathbf{e}.
\end{aligned}$$

† In (16), it is possible to decompose Φ_t into

$$\Phi_t^0 + \begin{bmatrix} (1-\tau)S/a_t \\ 0 \end{bmatrix} [r_t I, w_t I],$$

where Φ_t^0 is a lower block triangular matrix. This type of interaction is a generalization of a dyad we have in Example 2.1.

Equation (15) describes the way that the average capital stock of a household evolves with time, where ξ_t represents an interaction between the average and the covariance between the individual households' propensity to save and their individual employment levels and capital stocks. Equation (16) describes the way that variables deviate from the averages.

2.1. Steady State

Denote steady-state values by the subscript ∞. In the steady state $a_\infty = 1$ from (14), and k_∞ is determined from (15) by

$$\delta k_\infty = \bar{s}[\tau\beta + (1 - \tau)]f(k_\infty) + \xi_\infty. \tag{17}$$

Later, we shall need the expression for $f(k_\infty)/k_\infty$. It is given by

$$f(k_\infty)/k_\infty = (\delta - \xi_\infty/k_\infty)/\bar{s}[\tau\beta + (1 - \tau)]. \tag{18}$$

Under the condition that $I - \Phi_\infty$ is invertible we have

$$\begin{pmatrix} \kappa_\infty \\ \eta_\infty \end{pmatrix} = \phi_\infty(I - \Phi_\infty)^{-1}\begin{pmatrix} \mathbf{s} - \bar{s}\mathbf{e} \\ 0 \end{pmatrix}.$$

We examine κ_∞ since it turns out that

$$\eta_\infty = [\delta_2 r_\infty/(1 - \gamma_2 - w_\infty \delta_2)]\kappa_\infty. \tag{19}$$

From (18), we have, after scaling it by k_∞,

$$\kappa_\infty/k_\infty = u(I + v\Sigma)^{-1}\Sigma\mathbf{e}, \tag{20}$$

where $\Sigma = S - \bar{s}I$ and

$$u = \frac{(1 - \tau + \tau\beta)[f(k_\infty)/k_\infty]}{\delta(1 - \nu\bar{s})},$$

$$v = -\nu/(1 + \bar{s}\nu),$$

where

$$\nu = (1 - \tau)r_\infty\gamma_2/[(1 - \delta)(\gamma_2 + w_\infty \delta_2)].$$

We can write $u = \theta(1 - \bar{s}v)$, where

$$\theta = \frac{1}{\bar{s}}\left(\frac{\delta - \xi_\infty/k_\infty}{1 - \delta}\right). \tag{20$'$}$$

Note that $0 < v < 1$. From (20)

$$\begin{aligned} \mathrm{cov}^1_\infty/k_\infty &= \mathbf{e}'\Sigma(\kappa_\infty/k_\infty)/M \\ &= u\mathbf{e}'\Sigma(I + v\Sigma)^{-1}\Sigma\mathbf{e}/M = \theta(1 - \bar{s}v)\mathbf{e}'\Sigma(I + v\Sigma)^{-1}\Sigma\mathbf{e}/M. \end{aligned} \tag{21}$$

From (19) and (20)

$$\mathrm{cov}^2_\infty = -[\delta_2 r_\infty/(\gamma_2 + w_\infty \delta_2)]\,\mathrm{cov}^1_\infty.$$

Substituting the preceding two expressions in ξ_∞, we can write it as

$$\xi_\infty/k_\infty = \nu(\mathrm{cov}^1_\infty/k_\infty). \tag{22}$$

Substitute (21) into (22) and solve the resultant expression for ξ_∞/k_∞. It is given by

$$\xi_\infty/k_\infty = 1 - \varepsilon, \tag{23}$$

where ε is between 0 and 1 and is given by

$$\varepsilon = 1/(1 - vE/\bar{s}),$$

where

$$E = \mathbf{e}'\Sigma(I + v\Sigma)^{-1}\Sigma\mathbf{e}.$$

Substituting (23) into (18), and noting that $f(k_\infty) = (1 - \gamma)/\delta$ from (14), we can express k_∞ as

$$k_\infty = \left[\bar{s}[\tau\beta + (1 - \beta)]\frac{(1 - \gamma)}{\delta^2}\right]\left[1 + \frac{v}{1 - \tau}E\right]. \tag{23$'$}$$

The second term depending on E in (24) expresses the effects of nonidentical propensity to save on k_∞. Substituting (23′) into θ in (20′) gives

$$\theta = \frac{1 - \tau(1 - \beta)}{\bar{s}[\tau\beta + (1 - \beta)]}\frac{\delta\varepsilon}{(1 - \sigma)}. \tag{24}$$

Rewrite u as

$$u = \hat{u}\varepsilon, \tag{25}$$

where

$$\hat{u} = \frac{1 - \tau(1 - \beta)}{\bar{s}[\tau\beta + (1 - \beta)]}\frac{\delta}{(1 - \delta)(1 + \bar{s}\nu)} = \frac{\alpha\delta}{(1 - \delta)(1 + \bar{s}\nu)}. \tag{26}$$

We take as a measure of variance the coefficient of variation μ defined by

$$\mu^2 = \boldsymbol{\kappa}'_\infty\boldsymbol{\kappa}_\infty/k^2_\infty M = \hat{u}^2 g(v)\varepsilon^2/M, \tag{27}$$

where

$$g(v) = \mathbf{e}'\Sigma(I + v\Sigma)^{-2}\Sigma\mathbf{e} = \sum_i \frac{\sigma_i^2}{(1 + v\sigma_i)^2}.$$

Equations (20) and (25) show that κ^i_∞ is proportional to $\sigma_i(1 + v\sigma_i)$, and the constant of proportionality is determined by two factors, one depending on macro instruments and the other depending on the joint micro variables, since

$$\kappa^i_\infty/k_\infty = \hat{u}\left(1 + \alpha v\sum_j \frac{\sigma_j^2}{1 + v\sigma_j}\right)^{-1}\sigma_i\Big/(1 + v\sigma_i).$$

Equation (27) shows that μ is influenced by two factors. One represents the effect due to macro variables that affect $\hat{u}$ and v, the other the effect of the σ's, i.e., deviations of the s's from $\bar{s}$ which change ε and $g(v)$. For small v (i.e., for large $\bar{s}(1 - \gamma)/\delta^2$)

$$g(v)\varepsilon^2 = \frac{\mathbf{e}'\Sigma(I + v\Sigma)^{-2}\Sigma\mathbf{e}}{1 + \alpha v\mathbf{e}'\Sigma(I + v)^{-1}\Sigma\mathbf{e}^2}$$
$$\simeq \mathbf{e}'\Sigma^2\mathbf{e}[1 - 2v(\mathbf{e}'\Sigma^3\mathbf{e}/\mathbf{e}'\Sigma^2\mathbf{e} + \alpha\mathbf{e}'\Sigma^2\mathbf{e})].$$

Thus, to the zeroth order of approximation

$$\boldsymbol{\kappa}_\infty/k_\infty = \hat{u}\Sigma\mathbf{e}$$

and

$$\mu^2 \simeq \hat{u}^2\mathbf{e}'\Sigma^2\mathbf{e}/M.$$

In words, the relative shape of the spread of $\boldsymbol{\kappa}_\infty$ is entirely determined by $\Sigma\mathbf{e}$, the size being determined by $\hat{u}$. For small v (i.e., for large k_∞), $\partial\hat{u}/\partial\tau < 0$ and $\partial\hat{u}/\partial\beta < 0$ for small τ and β i.e., the higher the tax rate and the basic deduction, the smaller the absolute size of the spread, as well as the coefficient of variation.

To the first order of approximation

$$\boldsymbol{\kappa}_\infty/k_\infty = \frac{\hat{u}}{(1 + \alpha vE)}(I + v\Sigma)^{-1}\Sigma\mathbf{e} = \hat{u}\{[1 - v(\mathbf{e}'\Sigma^2\mathbf{e})]\Sigma\mathbf{e} - v\Sigma^2\mathbf{e}\} \tag{28}$$

and

$$\mu^2 \simeq \hat{u}^2\mathbf{e}'\Sigma^2\mathbf{e}[1 - 2v(\mathbf{e}'\Sigma^3\mathbf{e}/\mathbf{e}'\Sigma^2\mathbf{e} + \alpha\mathbf{e}'\Sigma^2\mathbf{e})]/M. \tag{29}$$

Now (28) shows that there are two components $\Sigma\mathbf{e}$ and $\Sigma^2\mathbf{e}$ in the distribution of $\boldsymbol{\kappa}_\infty$, v being the relative weight of the second vector.

Using the coefficient of skewness defined as $g = M^{1/2}\mathbf{e}'\Sigma^3\mathbf{e}/(\mathbf{e}'\Sigma^2\mathbf{e})^{3/2}$, (29) can be rewritten as

$$\mu = d[1 - 2v(gd + \bar{s} - d^2/\bar{s})]^{1/2}/s,$$

where $d = (\mathbf{e}'\Sigma^2\mathbf{e}/M)^{1/2}$.

2.2. Comparative Statics

Since

$$\frac{1}{\mu}\frac{\partial\mu}{\partial\tau} = \frac{1}{\hat{u}}\frac{\partial\hat{u}}{\partial\tau} - \frac{v\,d(d + \alpha M^{1/2}\,d)}{1 - 2v\,d(g + \alpha M^{1/2}\,d)} \times \left(\frac{1}{v}\frac{\partial v}{\partial\tau} + \frac{M^{1/2}\,d}{g + \alpha M^{1/2}\,d}\frac{\partial x}{\partial\tau}\right),$$

where

$$\partial \hat{u}/\partial \tau < 0, \qquad \partial \nu/\partial \tau < 0, \qquad \partial \alpha/\partial \tau < 0,$$

we see that τ has the opposing effects on μ. If v is small and the first term in (29) dominates, then $\partial\mu/\partial\tau < 0$. Since $\partial v/\partial\beta = 0$ and from (26)

$$\frac{\partial \hat{u}}{\partial \beta} = \frac{\delta}{(1-\delta)(1+\bar{s}v)}\frac{\partial \alpha}{\partial \beta} > 0,$$

$$\frac{1}{\mu}\frac{\partial \mu}{\partial \beta} = \frac{1}{\alpha} - \frac{vM^{1/2}\,d}{1-2v\,d(g+\alpha M^{1/2}\,d)}\frac{\partial \alpha}{\partial \beta},$$

$$\partial\mu/\partial\beta \gtreqless 0 \qquad \text{if and only if}$$

$$1/\alpha \gtreqless vM^{1/2}\,d^2/[1-2v\,d(g+\alpha M^{1/2}\,d)]$$

or

$$v \gtreqless \bar{v},$$

where

$$\bar{v} = 1/[d(2g + 3\alpha M^{1/2}\,d)].$$

Thus, provided that v of (20) is less than $\bar{v}$ defined before, $\partial\mu/\partial\beta > 0$. Bur for v larger than $\bar{v}$, $\partial\mu/\partial\beta < 0$.

Example 2.3. We briefly refer to yet another example to support consideration of our prototype in the next section. This model is described in Section 7.2 of Aoki (Ref. 8). In this example there are n agents in n goods with varying preference parameters. Trades take place by money. The speed of price adjustment may also vary from good to good. There are thus at least two sources for different values of the parameters that specify the dynamics for price changes and changes in money stock held by traders.

3. Prototype Dynamic System

Following the examples of the previous section, we consider as a prototype a dynamic system described by

$$x_{t+1} = (F + \mathbf{gh}')\mathbf{x}_t + \mathbf{b}v_t \tag{30}$$

where v_t is a scalar and

$$F = \operatorname{diag}(f_1 \cdots f_n), \qquad \mathbf{g}' = (g_1 \cdots g_n),$$

$$\mathbf{h}' = (h_1 \cdots h_n), \qquad \mathbf{b} = (b_1 \cdots b_n).$$

The previous examples seem to indicate that the assumption of the identical subsystem dynamics simplifies the distributional question considerably. This is indeed the case with this prototype dynamics. The average defined by $m_t = \mathbf{e}'\mathbf{x}_t/n$ and the spread $\mathbf{s}_t = \mathbf{x}_t - m_t\mathbf{e}$ are governed by

$$m_{t+1} = \alpha m_t + \beta v_t + \mathbf{C}'\mathbf{s}_t, \tag{31}$$

where

$$\alpha = \mathbf{e}'F\mathbf{e}/n + (\mathbf{e}'\mathbf{g})(\mathbf{e}'\mathbf{h})/n,$$
$$\beta = \mathbf{e}'\mathbf{b}/n,$$
$$\mathbf{C}' = \mathbf{b}'/n + [(\mathbf{e}'\mathbf{g})/n]\mathbf{h}',$$

where $\mathbf{f}' = \mathbf{e}'F$ and

$$\mathbf{s}_{t+1} = \Phi\mathbf{s}_t + \boldsymbol{\zeta} m_t + (\mathbf{b} - \beta\mathbf{e})v_t, \tag{32}$$

where

$$\Phi = (I - \mathbf{e}\mathbf{e}'/n)(F + \mathbf{g}\mathbf{h}'),$$
$$\boldsymbol{\zeta} = (I - \mathbf{e}\mathbf{e}'/n)(\mathbf{f} + (\mathbf{e}'\mathbf{h})\mathbf{g}).$$

When the dynamics of the subsystems are identical, i.e., $F = \gamma I$, and the coupling effects are identical, i.e., $\mathbf{g} = \delta\mathbf{e}$ and $\mathbf{h} = \eta\mathbf{e}$ for some scalars γ, δ, and η, then (31) and (32) simplify to

$$m_{t+1} = \alpha m_t + \beta v_t,$$
$$\mathbf{s}_{t+1} = \gamma\mathbf{s}_t + (\mathbf{b} - \beta\mathbf{e})v_t, \tag{33}$$

where

$$\alpha = \gamma + n\,\delta\eta.$$

Note that the dynamics is entirely decoupled. The steady-state spread when it exists is entirely determined by $\mathbf{b} - \beta\mathbf{e}$, since it is given by

$$\mathbf{s}'_\infty\mathbf{s}_\infty/m_\infty^2 = [(1 - \alpha)/\beta]^2(\mathbf{b} - \beta\mathbf{e})'(\mathbf{b} - \beta\mathbf{e})/(1 - \gamma)^2. \tag{34}$$

Now suppose that $F = \gamma I$ and $\mathbf{h} = \eta\mathbf{e}$, but allow the elements of $\mathbf{g}$ to be different, i.e., the dynamics of the subsystem are identical but a common signal $\mathbf{e}'\mathbf{x}_t$ affects each subsystem differently. then the dynamics become

$$m_{t+1} = \alpha m_t + \beta v_t, \qquad \mathbf{s}_{t+1} = \gamma\mathbf{s}_t + \boldsymbol{\zeta}\mu_t + (\mathbf{b} - \beta\mathbf{e})v_t, \tag{35}$$

where

$$\alpha = \gamma + \eta(\mathbf{e}'\mathbf{g})/n.$$

Now m_t affects $\mathbf{s}_t$ dynamics. Since $\mathbf{g} = \boldsymbol{\xi} + \delta\mathbf{e}$ where $\mathbf{e}'\boldsymbol{\xi} = 0$, $\boldsymbol{\zeta} = (n\eta)\boldsymbol{\xi}$. Its effect is proportional to $\boldsymbol{\xi}$. Instead of (34), the steady-state spread when it exists is given by

$$\mathbf{s}_\infty'\mathbf{s}_\infty/m_\infty^2 = \frac{(1-\alpha)^2}{\beta^2(1-\gamma)^2}\left(\mathbf{b} - \beta\mathbf{e} + \frac{\beta}{1-\alpha}\boldsymbol{\zeta}\right)'\left(\mathbf{b} - \beta\mathbf{e} + \frac{\beta}{1-\alpha}\boldsymbol{\zeta}\right). \tag{34'}$$

One obtains an opposite situation, of $\mathbf{s}_t$ affecting the m-dynamics if we assume that $F = \gamma I$ and $\mathbf{g} = \delta\mathbf{e}$, i.e., the subsystem dynamics are identical and the effect of the aggregate signal is the same, but each subsystem contributes differently to the aggregate signal. Then the dynamics is given by

$$\begin{aligned} m_{t+1} &= \alpha m_t + \delta\mathbf{h}'\mathbf{s}_t + \beta v_t, \\ \mathbf{s}_{t+1} &= \gamma s_t + (\mathbf{b} - \beta\mathbf{e})v_t, \end{aligned} \tag{35'}$$

where writing $\mathbf{h} = \boldsymbol{\theta} + \eta\mathbf{e}$ with $\mathbf{e}'\boldsymbol{\theta} = 0$, the effect of $\mathbf{s}_t$ on m_{t+1} is proportional to $\boldsymbol{\theta}$. The expression for the steady-state spread now becomes

$$\mathbf{s}_\infty'\mathbf{s}_\infty/m_\infty^2 = \frac{(1-\alpha)^2}{(1-\gamma)^2}\frac{(\mathbf{b} - \beta\mathbf{e})'(\mathbf{b} - \beta\mathbf{e})}{(1+\Delta)^2}, \tag{34''}$$

where

$$\Delta = \frac{\delta}{(1-\gamma)\beta}\mathbf{h}'(\mathbf{b} - \beta\mathbf{e}).$$

Comparison with (34′) indicates that when $\mathbf{h}'(\mathbf{b} - \beta\mathbf{e}) > 0$, the system (35′) has a "smaller" coefficient of variation than the system (34) with $\boldsymbol{\zeta}'(\mathbf{b} - \beta\mathbf{e}) > 0$. So far we have assumed that the dynamics of the subsystem are identical. When we drop this assumption, the distributional problem becomes more complex. Consider a special case of (30) with

$$F = \gamma I + \Sigma,$$

where $\mathbf{e}'\Sigma\mathbf{e} = 0$, $\mathbf{h} = \eta\mathbf{e}$, and $\mathbf{g} = \delta\mathbf{e}$. Then (31) and (32) simplify somewhat to

$$m_{t+1} = \alpha m_t + (\boldsymbol{\sigma}'/n)\mathbf{s}_t + \beta v_t$$

and

$$\mathbf{s}_{t+1} = (\gamma I + \Sigma - \mathbf{e}\boldsymbol{\sigma}'/n)\mathbf{s}_t + \boldsymbol{\sigma} m_t + (\mathbf{b} - \beta\mathbf{e})v_t.$$

The steady-state solution, when it exists, is given by

$$m_\infty/v_\infty = (\beta + \Delta)/(1-\alpha)$$

and

$$\mathbf{s}_\infty = [(1-\gamma)I - \Sigma]^{-1}\left\{\frac{\beta+\Delta}{1-\alpha}\boldsymbol{\sigma} + [\mathbf{b} - \beta(1+\Delta)\mathbf{e}]\right\}v_\infty,$$

where

$$\Delta = \bar{\bar{\Delta}}/(1 - \bar{\bar{\Delta}}),$$

$$\bar{\bar{\Delta}} = \frac{\boldsymbol{\sigma}'}{n}[(1-\gamma)I - \Sigma]^{-1}\left[\frac{\beta}{1-\alpha}\boldsymbol{\sigma} + (\mathbf{b} - \beta\mathbf{e})\right],$$

$$\bar{\Delta} = \frac{\boldsymbol{\sigma}'}{n}[(1-\gamma)I - \Sigma]^{-1} = \left(\frac{\boldsymbol{\sigma}}{1-\alpha} - \mathbf{e}\right).$$

Hence

$$V = \mathbf{s}'_\infty \mathbf{s}_\infty / m_\infty^2 = \left(\boldsymbol{\sigma} + \frac{\mathbf{b} - \beta(1+\Delta)\mathbf{e}}{\mu_\infty/v_\infty}\right)' \times [(1-\gamma)I - \Sigma]^{-2}\left(\boldsymbol{\sigma} + \frac{\mathbf{b} - \beta(1+\Delta)\mathbf{e}}{\mu_\infty/v_\infty}\right). \quad (36)$$

Since the expression for V is complicated, we analyze a simple case for which we assume that Δ is small and we approximate V by

$$V \approx \left(\boldsymbol{\sigma} + \frac{1-\alpha}{\beta}\boldsymbol{\delta}\right)' H\left(\boldsymbol{\sigma} + \frac{1-\alpha}{\beta}\boldsymbol{\delta}\right),$$

where

$$\boldsymbol{\delta} = \mathbf{b} - \beta\mathbf{e}$$

and

$$H = [(1-\gamma)I - \Sigma]^{-2}.$$

Let $\mathbf{v}$ be an eigenvector corresponding to the smallest eigenvalue of H, i.e., $\min(1 - \gamma - \sigma_i)^{-2}$. For example, if $(1 - \gamma - \sigma_1)^{-2}$ is minimal, then $\mathbf{v} = \mathbf{e}_1$. Then choosing $\boldsymbol{\delta}$ to be $(\mathbf{v} - \boldsymbol{\sigma})\beta/(1 - \alpha)$ will minimize V with respect to $\boldsymbol{\delta}$. $\boldsymbol{\delta}$ will not be zero in general. This would be appropriate if $\mathbf{b}$ (hence $\boldsymbol{\delta}$) is under the control of the central decision makers.

We have, with this choice of $\boldsymbol{\delta}$,

$$\frac{\partial V}{\partial \beta} = -2\frac{(1-\alpha)}{\beta^2}\boldsymbol{\delta}' H\left(\boldsymbol{\sigma} + \frac{1-\alpha}{\beta}\boldsymbol{\delta}\right) = -\frac{2}{\beta}\frac{1-\sigma_1}{(1-\gamma-\sigma_1)^2}.$$

Hence, if $\sigma_1 < 1$, then $\partial V/\partial\beta < 0$, and the larger β, the smaller V.

4. Discussion

We have suggested several dynamic systems composed of a number of "similar subsystems and derived equations that describe the time evolution of the average and the dispersion about the average of subsystem behavior. There are many other possibilities such as aggregate signals that weight subsystem states with statistical coefficients and so forth. Although it is not possible to examine these dynamics in detail in the short space allocated to this paper, I hope to have outlined an interesting class of dynamic systems that merits further examination.

Appendix: Calculation of Steady-State Solution

The steady-state solution is

$$\begin{pmatrix} m_\infty \\ \mathbf{s}_\infty \end{pmatrix} = \left[A - \frac{\delta}{n}\begin{pmatrix} 1 \\ -\mathbf{e} \end{pmatrix}(0, \boldsymbol{\zeta}')\right]^{-1}\left\{-\delta\begin{pmatrix} \bar{d} \\ \mathbf{v} \end{pmatrix} + (\alpha - a)\begin{pmatrix} \bar{\gamma} \\ \boldsymbol{\zeta} \end{pmatrix} - \begin{pmatrix} 0 \\ \bar{\gamma}\boldsymbol{\varepsilon} + \boldsymbol{\theta} \end{pmatrix}\right\},$$

where

$$A = I - \Phi^0.$$

Since A is block upper triangular, its inverse is easily calculated to be

$$A^{-1} = \begin{pmatrix} A^{11} & 0 \\ A^{21} & A^{22} \end{pmatrix},$$

where

$$\begin{aligned} A^{11} &= [1 + (n - 1)\,\delta\bar{\gamma}]^{-1}, \\ A^{21} &= -(I - \delta\Gamma)^{-1}\,\delta(n - 1)[1 + (n - 1)\,\delta\bar{\gamma}]^{-1}\boldsymbol{\zeta}, \\ A^{22} &= (I - \delta\Gamma)^{-1}. \end{aligned}$$

Note that

$$\begin{aligned} (I - \Phi)^{-1} &= \left[A - \frac{\delta}{n}\begin{pmatrix} 1 \\ -e \end{pmatrix}(0, \boldsymbol{\zeta}')\right]^{-1} \\ &= A^{-1} + \frac{\delta}{n}A^{-1}\begin{pmatrix} 1 \\ -\mathbf{e} \end{pmatrix}(0, \boldsymbol{\zeta}')A^{-1}/(1 - \Delta), \end{aligned}$$

where

$$\begin{aligned} \Delta &= \frac{\delta}{n}(0, \boldsymbol{\zeta}')A^{-1}\begin{pmatrix} 1 \\ -\mathbf{e} \end{pmatrix} \\ &= \frac{\delta}{n}\boldsymbol{\zeta}'(A^{21} - A^{22}\mathbf{e}) \\ &= -\frac{\delta}{n}\left\{\frac{\delta(n - 1)}{1 + \delta(n - 1)\bar{\gamma}}\sum_i\left(\frac{\zeta_i^2}{1 - \delta\gamma - \delta\zeta_i}\right) + \sum_i\left(\frac{\zeta_i}{1 - \delta\gamma - \delta\zeta_i}\right)\right\}. \end{aligned}$$

In the case where $\bar{d} = 0$,

$$m_\infty = \frac{\left\{(\alpha - a)\bar{\gamma} + \dfrac{\delta/n}{1-\Delta} U\right\}}{[1 + (n-1)\delta\bar{\gamma}]},$$

where U is defined by (5), and

$$\begin{aligned}
\mathbf{s}_\infty &= -\delta A^{22}\mathbf{v} + (\alpha - a)(A^{21}\bar{\gamma} + A^{22}\boldsymbol{\zeta}) - A^{22}(\bar{\gamma}\boldsymbol{\varepsilon} + \boldsymbol{\theta}) \\
&\quad + \frac{\delta/n}{1-\Delta}(A^{21} - A^{22}\mathbf{e}) \\
&\qquad \times\{-\delta\boldsymbol{\zeta}'A^{22}\mathbf{v} + (\alpha - a)\boldsymbol{\zeta}'(A^{21}\bar{\gamma} + A^{22}\boldsymbol{\zeta}) - \boldsymbol{\zeta}'A^{22}(\bar{\gamma}\boldsymbol{\varepsilon} + \boldsymbol{\theta})\} \\
&= \{-\delta(n-1)m_\infty\}(I - \delta\Gamma)^{-1}\boldsymbol{\zeta} + A^{22}\boldsymbol{\nu},
\end{aligned}$$

where

$$\boldsymbol{\nu} = -\delta\mathbf{v} + (\alpha - a)\boldsymbol{\zeta} - \bar{\gamma}\boldsymbol{\varepsilon} - \boldsymbol{\theta} - \frac{\delta/n}{1-\Delta}\tau\mathbf{e},$$

and where

$$\begin{aligned}
\tau &= -\delta\boldsymbol{\zeta}'A^{22}\mathbf{u} + (\alpha - a)\boldsymbol{\zeta}'(A^{21}\bar{\gamma} + A^{22}\boldsymbol{\zeta}) - \boldsymbol{\zeta}'A^{22}(\bar{\gamma}\boldsymbol{\varepsilon} + \boldsymbol{\theta}) \\
&= \frac{(\alpha - a)}{1 + (n-1)\,\delta\bar{\gamma}} V - \delta U - \boldsymbol{\zeta}'(I - \delta\Gamma)^{-1}(\gamma\boldsymbol{\varepsilon} + \boldsymbol{\theta}).
\end{aligned}$$

Equation (6) shows that $\mathbf{s}_\infty$ has a component determined by $\boldsymbol{\zeta}$ which is interdependent on $\boldsymbol{\zeta}$, $\mathbf{u}$, and other parameter variations $\boldsymbol{\varepsilon}$ and $\boldsymbol{\theta}$.

Let $\boldsymbol{\nu}' = \boldsymbol{\nu} - (\alpha - a)\boldsymbol{\zeta}$. Then

$$\begin{aligned}
A_{22}\boldsymbol{\nu}' &= -\delta(I - \delta\Gamma)^{-1}\mathbf{v} - (I - \delta\Gamma)^{-1}(\gamma\boldsymbol{\varepsilon} + \boldsymbol{\theta}) \\
&\quad - \frac{(\delta/n)\tau}{1-\Delta}(I - \delta\Gamma)^{-1}\mathbf{e}.
\end{aligned}$$

We can write m_∞ to reveal its dependence on $\mathbf{v}_\infty$ and $\boldsymbol{\zeta}$ by writing it as

$$m_\infty = m_\infty^r - \frac{\delta}{n(1-\Delta)[1 + (n-1)\,\delta\bar{\gamma}]} U,$$

where

$$m_\infty' = \frac{(\alpha - a)\bar{\gamma}}{[1 + (n-1)\,\delta\bar{\gamma}]}.$$

If V and U are small,

$$\frac{\mathbf{s}_\infty}{\bar{q}_\infty} = -\frac{\delta(n-1)(1+\kappa)}{(I-\delta\Gamma)^{-1}\boldsymbol{\zeta}} + \left(\frac{\alpha - a}{m_\infty}\right)(I-\delta\Gamma)^{-1}\boldsymbol{\zeta} + \frac{A^{22}\boldsymbol{\nu}^*}{m_\infty},$$

where

$$\begin{aligned} A^{22}\boldsymbol{\nu}^* &= -\delta(I-\delta\Gamma)^{-1}\mathbf{v} - (I-\delta\Gamma)^{-1}(\bar{\gamma}\boldsymbol{\varepsilon} + \boldsymbol{\theta}) \\ &\quad - \frac{(\delta/n)\tau}{1-\Delta}(I-\sigma\Gamma)^{-1}\mathbf{e} \\ &\simeq -\delta(I-\delta\Gamma)^{-1}\mathbf{v} \end{aligned}$$

if V and U are small and the variations in the γ's and α's are uncorrelated. Approximately, then,

$$\frac{\mathbf{s}_\infty}{m_\infty} = \frac{-\delta(n-1)(1+\kappa)+(\alpha-a)/m_\infty}{(I-\delta\Gamma)^{-1}\boldsymbol{\zeta}} + \frac{A^{22}\nu^*}{m_\infty}.$$

References

1. Aoki, M., On fluctuations in microscopic states of a large system, *Directions in Large-Scale Systems.* Edited by Y. C. Ho and S. K. Mitter, Plenum Press, New York, 1976.
2. Liggett, T. H., The Stochastic Evolution of Infinite Systems of Interacting Particles, No. 598, Lecture Notes in Mathematics, Springer-Verlag, New York, 1977.
3. Aoki, M., and Li, M. T., Partial reconstruction of state vectors in decentralized dynamic systems, *IEEE Transactions Automatic Control,* Vol. AC-18, pp. 289–292, 1973.
4. Stiglitz, J. E., Distribution of income and wealth among individuals, *Econometrica,* Vol. 37, pp. 382–397, 1969.
5. Pryor, F. L., Simulation of the impact of social and economic institutes on the size distribution of income and wealth, *American Economic Review,* Vol. 63, pp. 50–72, 1973.
6. Conlisk, J., An exploratory model of the size distribution of income, *Economic Inquiry,* Vol. 15, pp. 345–366, 1977.
7. Fukao, T., *Sisutem No Su-uri* (Mathematics of Systems), Chikuma Shobo, Tokyo, 1975.
8. Aoki, M., *Optimal Control and System Theory in Dynamic Economic Analysis,* North-Holland, New York, 1976.
9. Aoki, M., Large decentralized decision problems with imperfect information, *A link Between Science and Applications of Automatic Control,* Edited by A. Niemi, Pergamon Press, Oxford/New York, pp. 829–832, 1979.

Supplementary References

Aoki, M., Local controllability of a decentralized economic system, *Reviews of Economic Studies,* Vol. 41, pp. 51–63, 1974.

AOKI, M., Interaction among economic agents under imperfect information: An example, *New Trends in Dynamic System Theory and Economics.* Edited by A. Vlanzollo and M. Avler, Academic Press, New York, 1978.

BLINDER, A. S., *Toward an Economic Theory of Income Distribution*, MIT Press, Cambridge, Mass., 1974.

LUCAS, R. E., An equilibrium model of the business cycle, *Journal of Political Economy*, Vol. 83, pp. 1113–1144, 1975.

12

Hierarchical Decisionmaking under Uncertainty

TAMER BAŞAR

1. Introduction

In this paper we deal with two specific information structures within the context of hierarchical noncooperative equilibria of two-person, two-criteria quadratic Gaussian decision problems. Specifically, let x denote the unknown state of nature, which is an n-dimensional Gaussian random vector with mean zero and covariance Σ. Furthermore, let u_i denote the decision variable of decisionmaker (DM) $i, i = 1, 2$, which takes values in $\mathscr{R}^{r_i}$. Then the quadratic objective functionals of DM1 and DM2 are, respectively, defined by

$$\begin{aligned} g_1(x, u_1, u_2) = {} & \tfrac{1}{2}u_1^T D_{11} u_1 + u_1^T D_{12} u_2 \\ & + \tfrac{1}{2}u_2^T D_{13} u_2 + u_1^T C_{11} x + u_2^T C_{12} x + u_1^T k_{11} + u_2^T k_{12}, \end{aligned} \tag{1a}$$

$$\begin{aligned} g_2(x, u_1, u_2) = {} & \tfrac{1}{2}u_2^T D_{22} u_2 + u_2^T D_{21} u_1 \\ & + \tfrac{1}{2}u_1^T D_{23} u_1 + u_1^T C_{21} x + u_2^T C_{22} x + u_1^T k_{21} + u_2^T k_{22}, \end{aligned} \tag{1b}$$

where $D_{ij}, C_{ik}, i, k = 1, 2, j = 1, 2, 3$, are constant matrices of appropriate dimensions with $D_{ii} > 0$, and where k_{ij} are appropriate dimensional constant vectors.

The hierarchical noncooperative equilibrium solution concept to be considered in this paper allows one of the decisionmakers to have access

TAMER BAŞAR • Applied Mathematics Division, Marmara Research Institute, Gebze, Kocaeli, Turkey.

to the strategy of the other decisionmaker. This is, in essence, the Stackelberg equilibrium solution concept of game theory. Under a given information structure (which will be made precise later), let γ_i denote an admissible strategy† of DMi. Then we will adopt the convention that DM1 announces his strategy first, and DM2 reacts accordingly. For each admissible γ_1, DM2 will pick a decision rule γ_2 to minimize the expected value of g_2. If this is repeated for all admissible γ_1 (and under a given information structure), then we obtain the reaction set of DM2. Minimization of the expected value of g_1 over that reaction set yields the optimal decision rule of DM1 under the hierarchical noncooperative equilibrium solution concept.

The two information structures to be considered in this paper will respectively be of static and dynamic nature. To delineate them, let z_i be an m_i-dimensional Gaussian random vector defined by

$$z_i = H_i x + w_i, \qquad i = 1, 2, \tag{2}$$

where H_i is a constant matrix of appropriate dimensions and w_i is an m_i-dimensional Gaussian random vector with mean zero and covariance $\Lambda_i > 0$. Then if η_i denotes the information available to DMi, we consider, in the sequel, the following two information structures:

A. *Static Information Structure.* $\eta_i = \{z_i\}$, $i = 1, 2$.
B. *Dynamic Information Structure.* $\eta_1 = \{z_1, z_2, u_2\}$, $\eta_2 = \{z_2\}$.

It should be noted that the dynamic information structure also provides DM1 with the observation (z_2) and the decision value (u_2) of DM2, and it is a special case of what is known as the *one-step-delay information sharing pattern* in the literature (Ref. 8). Such an information structure was also considered in Ref. 3, but within the context of Nash equilibria.

In the sequel, we shall verify that the decision problem just formulated admits unique Stackelberg equilibria under both information structures, provided that certain convexity conditions are satisfied. Furthermore, it will be shown that the equilibrium decision rules are linear, in each case, and their exact expressions will be given. Since the methods of proof and derivation are quite different under static and dynamic information, they will be treated separately, with Section 2 devoted to the former and Section 3 to the latter. In Section 4, we include some applications of the results of Section 2 in microeconomics in the determination of Stackelberg equilibria for both the quantity and price models of duopoly. The paper ends with concluding remarks.

† An admissible strategy is a measurable mapping from an appropriate information space into the decision space and satisfies a finite norm property. For a mathematically precise definition, the reader is referred to Ref. 1.

2. Hierarchical Equilibria under Static Information

Under static information and in the light of the hierarchical noncooperative equilibrium solution concept introduced in Section 1, we first determine the reaction strategy of DM2 to an arbitrarily selected admissible decision rule γ_1 of DM1. To this end, we minimize the expected value of g_2, with $u_1 = \gamma_1(z_1)$ and conditioned on the observed value of z_2. The result is the unique relation

$$\gamma_2(z_2) = -D_{22}^{-1}C_{22}E[x \mid z_2] - D_{22}^{-1}k_{22} - D_{22}^{-1}D_{21}E[\gamma_1(z_1) \mid z_2], \quad (3)$$

where $E[\cdot \mid z_2]$ denotes the expected value of $(\cdot)$ conditioned on the observed value of z_2. Thus (3) constitutes the unique reaction of DM2 to an announced strategy of DM1. To determine the hierarchical equilibrium strategy of DM1, we have to minimize the expected value of g_1, with $u_2 = \gamma_2(z_2)$ as given by (3) and conditioned on the observed value of z_1. Substitution of (3) into (1a) with $u_2 = \gamma_2(z_2)$ and taking the expectation yields,

$$\begin{aligned} J(\gamma_1) = E[&\tfrac{1}{2}\gamma_1^T D_{11}\gamma_1 + \gamma_1^T C_{11}x - \gamma_1^T D_{12}D_{22}^{-1}C_{22}E[x \mid z_2] \\ &- \gamma_1^T D_{12}D_{22}^{-1}k_{22} \\ &- \gamma_1^T D_{12}D_{22}^{-1}D_{21}E[\gamma_1 \mid z_2] - E[x^T \mid z_2]C_{22}^T D_{22}^{-1}C_{12}x \\ &+ \gamma_1^T k_{11} - k_{22}^T D_{22}^{-1}C_{12}x - E[\gamma_1^T \mid z_2]D_{21}^T D_{22}^{-1}C_{12}x \\ &+ \tfrac{1}{2}E[\gamma_1^T \mid z_2]D_{21}^T D_{22}^{-1}D_{13}D_{22}^{-1}D_{21}E[\gamma_1 \mid z_2] \\ &+ k_{22}^T D_{22}^{-1}D_{13}D_{22}^{-1}C_{22}E[x \mid z_2] \\ &+ \tfrac{1}{2}E[x^T \mid z_2]C_{22}^T D_{22}^{-1}D_{13}D_{22}^{-1}C_{22}E[x \mid z_2] \\ &+ k_{22}^T D_{22}^{-1}D_{13}D_{22}^{-1}D_{21}E[\gamma_1 \mid z_2] \\ &+ E[\gamma_1^T \mid z_2]D_{21}^T D_{22}^{-1}D_{13}D_{22}^{-1}C_{22}E[x \mid z_2] \\ &+ \tfrac{1}{2}k_{22}^T D_{22}^{-1}D_{13}D_{22}^{-1}k_{22} - k_{12}^T D_{22}^{-1}k_{22} \\ &- k_{12}^T D_{22}^{-1}(C_{22}E[x \mid z_2] + D_{21}E[\gamma_1 \mid z_2])], \end{aligned} \quad (4)$$

where we have suppressed the z_1-dependence of γ_1. Even though (4) is a homogeneous function of γ_1 of degree two, its minimization is not a standard one. We first note that if $\hat{\gamma}_1$ provides a minimum value to J, then any admissible linear variation around $\hat{\gamma}_1$ belongs to the same linear space as $\hat{\gamma}_1$ does. Furthermore, for any admissible linear variation h around $\hat{\gamma}_1$, we can express $J(\hat{\gamma}_1 + h)$ as

$$J(\hat{\gamma}_1 + h) = J(\hat{\gamma}_1) + \delta J(\hat{\gamma}_1; h) + \tfrac{1}{2}\delta^2 J(\hat{\gamma}_1; h), \quad (5)$$

where δJ and $\delta^2 J$ are, respectively, the first and second Gateaux variations of J around $\hat{\gamma}_1$.† The first Gateaux variation $\delta J(\hat{\gamma}_1; h)$ is defined by

$$\delta J(\hat{\gamma}_1; h) \triangleq \lim_{\alpha \to 0} \frac{d}{d\alpha} J(\hat{\gamma}_1 + \alpha h),$$

which in our case is

$$\begin{aligned}
\delta J(\hat{\gamma}_1; h) = {} & E[h^T D_{11} \hat{\gamma}_1 + h^T C_{11} x - h^T D_{12} D_{22}^{-1} C_{22} E[x \mid z_2] \\
& - h^T D_{12} D_{22}^{-1} k_{22} \\
& - h^T D_{12} D_{22}^{-1} D_{21} E[\hat{\gamma}_1 \mid z_2] - \gamma_1^T D_{12} D_{22}^{-1} D_{21} E[h \mid z_2] \\
& + h^T k_{11} + k_{22}^T D_{22}^{-1} D_{13} D_{22}^{-1} D_{21} E[h \mid z_2] \\
& \qquad - E[h^T \mid z_2] D_{21}^T D_{22}^{-1} C_{12} x \\
& + E[h^T \mid z_2] D_{21}^T D_{22}^{-1} D_{13} D_{22}^{-1} D_{21} E[\hat{\gamma}_1 \mid z_2] \\
& \qquad - k_{12}^T D_{22}^{-1} D_{21} E[h \mid z_2] \\
& + E[h^T \mid z_2] D_{21}^T D_{22}^{-1} D_{13} D_{22}^{-1} C_{22} E[x \mid z_2]],
\end{aligned} \tag{6}$$

where we have again suppressed the z_1-dependence of $\hat{\gamma}_1$ and h. Since "expectation is expectation of conditional expectation," the following relation holds for all vector functions k:

$$\begin{aligned}
E[E[h^T \mid z_2] k(x, z_1)] &= E[E[h^T \mid z_2] E[k(x, z_1) \mid z_2]] \\
&= E[h^T E[k(x, z_1) \mid z_2]] \\
&= E[E[E[k^T(x, z_1) \mid z_2] h \mid z_1]].
\end{aligned}$$

If this relation is used in (6), the first Gateaux variation can be expressed as

$$\delta J(\hat{\gamma}_1; h) = E[h^T(z_1) l(z_1, \hat{\gamma}_1)], \tag{7a}$$

where

$$\begin{aligned}
l(z_1, \hat{\gamma}_1) = {} & D_{11} \hat{\gamma}_1(z_1) + C_{11} E[x \mid z_1] - D_{12} D_{22}^{-1} C_{22} E[E[x \mid z_2] \mid z_1] \\
& - D_{12} D_{22}^{-1} k_{22} - (D_{12} D_{22}^{-1} D_{21} \\
& + D_{21}^T D_{22}^{-1} D_{12}^T) E[E[\hat{\gamma}_1 \mid z_2] \mid z_1] \\
& - D_{21}^T D_{22}^{-1} C_{12} E[E[x \mid z_2] \mid z_1] + k_{11} \\
& + D_{21}^T D_{22}^{-1} D_{13} D_{22}^{-1} k_{22} - D_{21}^T D_{22}^{-1} k_{12} \\
& + D_{21}^T D_{22}^{-1} D_{13} D_{22}^{-1} D_{21} E[E[\hat{\gamma}_1 \mid z_2] \mid z_1] \\
& + D_{21}^T D_{22}^{-1} D_{13} D_{22}^{-1} C_{22} E[E[x \mid z_2] \mid z_1].
\end{aligned} \tag{7b}$$

The second Gateaux variation, on the other hand, is defined by

$$\delta^2 J(\hat{\gamma}_1; h) \triangleq \lim_{\alpha \to 0} \frac{d^2}{d\alpha^2} J(\hat{\gamma}_1 + \alpha h),$$

† Further details on mathematical justification of these steps can be found in (Ref. 2).

which in our case is

$$\delta^2 J(\hat{\gamma}_1; h) = E[h^T D_{11} h + E[h^T \mid z_2] D_{21}^T D_{22}^{-1} D_{13} D_{22}^{-1} D_{21} E[h \mid z_2] \\ - 2h^T D_{12} D_{22}^{-1} D_{21} E[h \mid z_2]].$$

This can equivalently be written as

$$\delta^2 J(\hat{\gamma}_1; h) = E[h^T D_{11} h + E[h^T \mid z_2](K - D_{11}) E[h \mid z_2]], \tag{8a}$$

where

$$K \triangleq D_{11} + D_{21}^T D_{22}^{-1} D_{13} D_{22}^{-1} D_{21} - D_{12} D_{22}^{-1} D_{21} - D_{21}^T D_{22}^{-1} D_{12}^T. \tag{8b}$$

Now, if $\hat{\gamma}_1$ provides a global minimum for $J(\gamma_1)$, we should have

$$J(\hat{\gamma}_1) \leq J(\hat{\gamma}_1 + h), \qquad \forall \text{ admissible } h,$$

and since h belongs to a linear class, it follows from (5) that a necessary and sufficient condition is

$$\delta J(\hat{\gamma}_1; h) = 0, \quad \delta^2 J(\hat{\gamma}_1; h) \geq 0, \qquad \forall \text{ admissible } h. \tag{9}$$

Using a contraction mapping principle, we can show that (9) admits a unique solution under the condition

$$0 < D_{11}^{-1/2} K D_{11}^{-1/2} < 2I \tag{10}$$

and, furthermore, that the unique solution is affine (Ref. 2). Hence, to obtain an expression for $\hat{\gamma}_1$, we substitute the affine structure $\hat{\gamma}_1(z_1) = Az_1 + q$ into (7b) and set it equal to zero. The resulting equation for A is

$$\begin{aligned} D_{11} A z_1 &+ C_{11} \Sigma H_1^T (H_1 \Sigma H_1^T + \Lambda_1)^{-1} z_1 \\ &- D_{12} D_{22}^{-1} C_{22} \Sigma H_2^T (H_2 \Sigma H_2^T + \Lambda_2)^{-1} H_2 \Sigma H_1^T (H_1 \Sigma H_1^T + \Lambda_1)^{-1} z_1 \\ &+ (K - D_{11}) A \Sigma H_2^T (H_2 \Sigma H_2^T + \Lambda_2)^{-1} H_2 \Sigma H_1^T (H_1 \Sigma H_1^T + \Lambda)^{-1} z_1 \\ &- D_{21}^T D_{22}^{-1} (C_{12} - D_{13} D_{22}^{-1} C_{22}) \Sigma H_2^T (H_2 \Sigma H_2^T + \Lambda_2)^{-1} H_2 \Sigma H_1^T (H_1 \Sigma H_1^T \\ &+ \Lambda_1)^{-1} z_1 = 0. \end{aligned}$$

This then yields the following Lyapunov-type equation from which A can be solved uniquely under (10):

$$D_{11} A + (K - D_{11}) A \Sigma_2 H_2 \Sigma_1 = [(D_{12} - D_{21}^T D_{22}^{-1} D_{13}) D_{22}^{-1} C_{22} \\ + D_{21}^T D_{22}^{-1} C_{12}] \Sigma_2 H_2 \Sigma_1 - C_{11} \Sigma_1, \tag{11a}$$

where

$$\Sigma_i \triangleq \Sigma H_i^T (H_i \Sigma H_i^T + \Lambda_i)^{-1}, \qquad i = 1, 2. \tag{11b}$$

Furthermore, the unique expression for q is

$$q = K^{-1}(D_{12}D_{22}^{-1}k_{22} + D_{21}^{T}D_{22}^{-1}k_{12} - k_{11} - D_{21}^{T}D_{22}^{-1}D_{13}D_{22}^{-1}k_{22}). \quad (11c)$$

This then completes the solution of the hierarchical decision problem under static information. We summarize the results of this section:

Theorem 2.1. The hierarchical noncooperative equilibrium solution of the decision problem with the static information structure is unique under condition (10), and it is given by

$$\hat{\gamma}_1(z_1) = Az_1 + q, \quad (12a)$$

$$\hat{\gamma}_2(z_2) = -D_{22}^{-1}(C_{22}\Sigma_2 z_2 + k_{22}) - D_{22}^{-1}D_{21}E[\hat{\gamma}_1(z_1) \mid z_2], \quad (12b)$$

where A is defined uniquely by (11a), q is given uniquely by (11c), and Σ_2 is defined by (11b).

3. Hierarchical Equilibria under Dynamic Information

Under dynamic information that provides DM1 with $\eta_1 = \{z_1, z_2, u_2\}$ and DM2 with $\eta_2 = \{z_2\}$, it is not possible to obtain the hierarchical noncooperative (Stackelberg) equilibria using standard techniques of optimization. The main reason for this is that it is, in general, not possible to obtain explicitly the reaction set of DM2, and this makes the optimization problem faced by DM1 a nonstandard one. To see this better let us introduce the notation

$$J_i(\gamma_1, \gamma_2) = E[g_i(x, u_1, u_2) \mid u_1 = \gamma_1(\eta_1), u_2 = \gamma_2(\eta_2)], \qquad i = 1, 2,\dagger \quad (13)$$

under the given dynamic information structure, and for $\gamma_1 \in \Gamma_1$, $\gamma_2 \in \Gamma_2$ where Γ_i is an appropriate linear space of Borel measurable mappings. Now, for each $\gamma_1 \in \Gamma_1$, DM2's reaction will be a $\gamma_2^0 \in \Gamma_2$ that satisfies the inequality

$$J_2(\gamma_1, \gamma_2^0) \leq J_2(\gamma_1, \gamma_2), \qquad \forall \gamma_2 \in \Gamma_2. \quad (14)$$

The strategies γ_2^0 thus obtained for each $\gamma_1 \in \Gamma_1$ form a reaction set $R(\gamma_1)$ in Γ_2. Since u_1 will be a function of u_2, the nature of $R(\gamma_1)$ will very much depend on the structure of γ_1. For example, if $\gamma_1(\eta_1)$ is taken to be linear in u_2, then $R(\gamma_1)$ will be a singleton for each γ_1 within that

† In this section, both g_1 and g_2 are considered without the terms $u_1^T k_{11}$, $u_2^T k_{12}$, $u_1^T k_{21}$, $u_2^T k_{22}$. This results in considerable simplification in some of the expressions involved while bringing about no loss of generality as far as structural and behavioral aspects of the solution are concerned.

class (because of the quadratic nature of g_2). However, for nonlinear $\gamma_1(\eta_1)$, $R(\gamma_1)$ will in general not be a singleton. Hence, to arrive at an optimal solution, DM1 will have to infimize the supremum of $J_1(\gamma_1, \gamma_2)$, where the supremum is taken over $R(\gamma_1)$ and the infimum over Γ_1. But since $R(\gamma_1)$ cannot explicitly be determined, this optimization problem is not solvable by any of the direct methods available. In the sequel, we will develop an indirect approach to solve the problem by relating the hierarchical decision problem under dynamic information to a particular team problem. To this end, let

$$J_1^* = \inf_{\Gamma_1} \sup_{R(\gamma_1)} J_1(\gamma_1, \gamma_2), \tag{15}$$

and note that a lower bound on J_1^* is

$$J_1^* \geq \min_{\gamma_1 \in \Gamma_1} \min_{\gamma_2 \in \Gamma_2} J_1(\gamma_1, \gamma_2) \triangleq J_1(\gamma_1^t, \gamma_2^t), \tag{16}$$

where the said minimizing solution exists if g_1 is convex in the pair (u_1, u_2). Hence, let us first make the following assumption.

Assumption 3.1. $D_{11} > 0$, $D_{13} - D_{12}^T D_{11}^{-1} D_{12} > 0$.

Then, the right-hand-side of (16) defines a well-defined team problem whose solution is given by the following lemma.

Lemma 3.1. Under Assumption 3.1 and the given dynamic information structure, a team solution for the right-hand side of (16) is given by

$$\gamma_1^t(\eta_1) = -D_{11}^{-1} D_{12} u_2 - D_{11}^{-1} C_{11}[K z_1 + (I - K H_1)\Sigma_2 z_2], \tag{17a}$$

$$\gamma_2^t(\eta_2) = -(D_{13} - D_{12}^T D_{11}^{-1} D_{12})^{-1}(C_{12} - D_{12}^T D_{11}^{-1} C_{11})\Sigma_2 z_2, \tag{17b}$$

where

$$K = P H_1^T (H_1 P H_1^T + \Lambda_1)^{-1}, \tag{18a}$$

$$P = \Sigma - \Sigma H_2^T \Sigma_2^T, \tag{18b}$$

and Σ_2 is defined by (11b).

Proof. First, minimizing $J_1(\gamma_1, \gamma_2)$ over Γ_1 for fixed $u_2 = \gamma_2(\eta_2)$, we obtain the unique expression

$$\gamma_1^t(\eta_1) = -D_{11}^{-1} D_{12} u_2 - D_{11}^{-1} C_{11} E[x \mid \eta_1],$$

and since $E[x \mid \eta_1] = E[x \mid z_1, z_2]$, (17a) follows by evaluation of the latter conditional expectation. Finally, substitution of (17a) into $J_1(\gamma_1, \gamma_2)$

and minimization of the resulting quadratic expression over Γ_2 yields (17b). □

Remark 3.1. Even though the pair of strategies (17) yields a global minimum for J_1 under the given dynamic information structure, (17a) is not the only element of Γ_1 that is optimal. Any strategy of the form

$$\gamma_1(\eta_1) = \gamma_1^t(\eta_1) + \varphi(u_2, \gamma_2^t(\eta_2)), \tag{19}$$

where φ is chosen such that $\varphi(u_2, u_2) = 0$, also yields a global minimum for J_1, together with γ_1^t given by (17b).

We now show that under certain conditions there exists a function φ such that (19) constitutes a Stackelberg strategy for DM1. In other words, there exists a φ such that if (19) is substituted into $J_2(\gamma_1, \gamma_2)$ and the resulting expression is minimized over $u_2 = \gamma_2(\eta_2) \in \Gamma_2$, the unique minimizing solution will be the team solution γ_2^t given by (17b).

To this end, we first note that the only possible structure for φ that could result in a linear $\gamma_2(\eta_2)$ after minimization is a linear structure. Thus, we take

$$\varphi(u_2, \gamma_2^t(\eta_2)) = S[u_2 - \gamma_2^t(\eta_2)], \tag{20}$$

where S is an appropriate dimensional matrix, and γ_2^t is as given by (17b). Now, if (19) is substituted into $J_2(\gamma_1, \gamma_2)$, together with (20), we obtain the following functional of γ_2:

$$\begin{aligned} F(\gamma_2) = E[&\tfrac{1}{2}u_2^T D_{22} u_2 - u_2^T D_{21}(D_{11}^{-1} D_{12} u_2 + D_{11}^{-1} C_{11} E[x \mid z_1, z_2] \\ &+ S[\gamma_2^t - u_2]) + \tfrac{1}{2}(D_{11}^{-1} D_{12} u_2 + D_{11}^{-1} C_{11} E[x \mid z_1, z_2] \\ &+ S[\gamma_2^t - u_2])^T D_{23}(D_{11}^{-1} D_{12} u_2 + D_{11}^{-1} C_{11} E[x \mid z_1, z_2] \\ &+ S[\gamma_2^t - u_2]) - (D_{11}^{-1} D_{12} u_2 + D_{11}^{-1} C_{11} E[x \mid z_1, z_2] \\ &+ S[\gamma_2^t - u_2])^T C_{21} x + u_2^T C_{22} x \mid u_2 = \gamma_2(\eta_2)]. \end{aligned} \tag{21}$$

This is a quadratic functional of u_2, and it admits a unique minimizing solution if and only if

$$\begin{aligned} D_{22} + D_{21} S + S^T D_{21}^T + (D_{11}^{-1} D_{12} - S)^T D_{23}(D_{11}^{-1} D_{12} - S) \\ - D_{21} D_{11}^{-1} D_{12} - D_{12}^T D_{11}^{-1} D_{21}^T > 0. \end{aligned} \tag{22}$$

This condition places a restriction on the class of matrices S that could be considered in the representation (20). Let us denote the class of all such matrices by Ω, i.e.,

$$\Omega \triangleq \{\text{all } (r_1 \times r_2)\text{-dimensional matrices } S \text{ that satisfy (22)}\}. \tag{23}$$

It should be noted that Ω is not an empty set, since if S is taken as $S = D_{11}^{-1} D_{12}$, then (22) becomes $D_{22} > 0$, which is true by hypothesis.

Let us now minimize $F(\gamma_2)$ with $S \in \Omega$ and with γ_2^t fixed as given by (17b). The first-order condition (which is also sufficient by strict convexity) is

$$\begin{aligned}
&D_{22}\hat{\gamma}_2 - D_{21}(D_{11}^{-1}D_{12}\hat{\gamma}_2 + D_{11}^{-1}C_{11}E[x \mid z_1] + S[\gamma_2^t - \hat{\gamma}_2]) \\
&- (D_{11}^{-1}D_{12} - S)^T D_{21}^T \hat{\gamma}_2 + (D_{11}^{-1}D_{12} - S)^T D_{23}(D_{11}^{-1}D_{12}\hat{\gamma}_2 \\
&\qquad\qquad + D_{11}^{-1}C_{11}E[x \mid z_2] \\
&+ S[\gamma_2^t - \hat{\gamma}_2]) - (D_{11}^{-1}D_{12} - S)^T C_{21}E[x \mid z_2] + C_{22}E[x \mid z_2] = 0, \quad (24)
\end{aligned}$$

where $\hat{\gamma}_2 = \hat{\gamma}_2(\eta_2)$ denotes the minimizing solution. In writing down (24) we have also made use of the following property of conditional expectation:

$$E[E[x \mid z_1, z_2] \mid \eta_2] = E[x \mid \eta_2] \equiv E[x \mid z_2],$$

since η_2 generates a coarser σ-field than (z_1, z_2).

We now seek to determine the conditions under which it is possible to find an $S \in \Omega$ such that (24) is satisfied by $\hat{\gamma}_2 = \gamma_2^t$ which is given by (17b). If we let $\hat{\gamma}_2 = \gamma_2^t$ in (24), we obtain

$$\begin{aligned}
&[D_{22} - D_{21}D_{11}^{-1}D_{12} + (D_{11}^{-1}D_{12} - S)^T(D_{23}D_{11}^{-1}D_{12} - D_{21}^T)]\gamma_2(\eta_2) \\
&- [D_{21}D_{11}^{-1}C_{11} + (D_{11}^{-1}D_{12} - S)^T(C_{21} - D_{23}D_{11}^{-1}C_{11}) - C_{22}]E[x \mid z_2] \\
&= 0,
\end{aligned}$$

and if we further make use of (17b), we obtain the condition

$$\begin{aligned}
&(D_{22} - D_{21}D_{11}^{-1}D_{12} + \tilde{S}^T(D_{23}D_{11}^{-1}D_{12} - D_{21}^T))(D_{13} - D_{12}^T D_{11}^{-1}D_{12})^{-1} \\
&\times (C_{12} - D_{12}^T D_{11}^{-1}C_{11}) + D_{21}D_{11}^{-1}C_{11} + \tilde{S}^T(C_{21} - D_{23}D_{11}^{-1}C_{11}) - C_{22} \\
&= 0,
\end{aligned}$$

where

$$\tilde{S} \triangleq D_{11}^{-1}D_{12} - S. \quad (25)$$

This can be written as

$$\tilde{A}\tilde{S} = \tilde{B}, \quad (26a)$$

where

$$\begin{aligned}
\tilde{A} &\triangleq C_{21}^T - C_{11}^T D_{11}^{-1}D_{23} + (C_{12}^T - C_{11}^T D_{11}^{-1}D_{12})(D_{13} - D_{12}^T D_{11}^{-1}D_{12})^{-1} \\
&\qquad \times (D_{12}^T D_{11}^{-1}D_{23} - D_{21}), \quad (26b) \\
\tilde{B} &\triangleq C_{22}^T - C_{11}^T D_{11}^{-1}D_{21}^T - (C_{12}^T - C_{11}^T D_{11}^{-1}D_{12})(D_{13} - D_{12}^T D_{11}^{-1}D_{12})^{-1} \\
&\qquad \times (D_{22} - D_{12}^T D_{11}^{-1}D_{21}^T). \quad (26c)
\end{aligned}$$

Furthermore, in terms of $\tilde{S}$, (22) can be rewritten as

$$D_{22} - D_{21}\tilde{S} - \tilde{S}^T D_{21}^T + \tilde{S}^T D_{23}\tilde{S} > 0. \quad (27)$$

Let us replace Ω by $\tilde{\Omega}$ which is defined by

$$\tilde{\Omega} \triangleq \{\text{all } (r_1 \times r_2)\text{-dimensional real matrices } \tilde{S} \text{ that satisfy (27)}\}. \quad (28)$$

Then we have the following result.

Theorem 3.1. If there exists an element $\tilde{S}^*$ of $\tilde{\Omega}$ that also satisfies (26a), the hierarchical noncooperative equilibrium solution of the decision problem with the dynamic information structure is linear, and the optimal strategies are given by

$$\begin{aligned}\hat{\gamma}_1(\eta_1) &= -\tilde{S}^* u_2 - D_{11}^{-1}C_{11}[Kz_1 + (I - KH_1)\Sigma_2 z_2] \\ &\quad + (\tilde{S}^* - D_{11}^{-1}D_{12})\hat{\gamma}_2(\eta_2), \end{aligned} \quad (29a)$$

$$\hat{\gamma}_2(\eta_2) = -(D_{13} - D_{12}^T D_{11}^{-1} D_{12})^{-1}(C_{12} - D_{12}^T D_{11}^{-1} C_{11})\Sigma_2 z_2. \quad (29b)$$

Proof. We have already shown that if (26a) admits a solution in $\tilde{\Omega}$, and if DM1 picks the strategy (29a), then (29b) is the unique response of DM2. However, (29b) is the team strategy γ_2^t given in Lemma 3.1, and $\hat{\gamma}_1(\eta_1)$ is a "representation" of the team strategy γ_1^t. Hence, in the light of Remark 3.1, it follows that the strategies (29) yield $J_1(\gamma_1^t, \gamma_2^t)$ which is the lower bound on J_1 defined by (15), and consequently they are in Stackelberg equilibrium. □

Remark 3.2. One property of the hierarchical noncooperative equilibrium solution presented here is that by picking (and announcing) the strategy (29a), DM1 can force DM2 to a team strategy that minimizes DM1's expected cost, while DM2 is in fact minimizing his own expected cost.

We now discuss existence of a solution to (26a) in the set $\tilde{\Omega}$. Let us first make the following assumption concerning the structure of g_2.

Assumption 3.2. $D_{22} > 0$, $D_{23} - D_{21}^T D_{22}^{-1} D_{21} > 0$.

Now since (27) can also be written as

$$(\tilde{S} - D_{23}^{-1}D_{21}^T)^T D_{23}(\tilde{S} - D_{23}^{-1}D_{21}^T) + D_{23} - D_{21}^T D_{22}^{-1} D_{21} > 0,$$

it follows that under Assumption 3.2 [which imposes a strict convexity restriction on g_2 in terms of (u_1, u_2)], $\tilde{\Omega}$ becomes the space of all $(r_1 \times r_2)$-dimensional matrices. Thus, the condition of Theorem 3.2 is reduced to the requirement that $\tilde{S}^*$ be any solution of the linear matrix equation (26a). In particular, if $r_1 = n$, then a sufficient condition is nonsingularity of the square matrix $\tilde{A}$.

If the preceding conditions are not satisfied, then one might arrive at the conclusion that a Stackelberg equilibrium does not exist. We will actually show in the sequel that such a conclusion is true for the case $r_1 \geq r_2$. It will be verified that there exists a sequence of strategies for DM1 that bring his expected cost arbitrarily close to the lower bound given in (16), but that sequence does not admit a well-defined limit. To see this, we again start with the strategy (19) for DM1, with φ given by (20), and substitute it into $J_1(\gamma_1, \gamma_2)$. The resulting expression is $F(\gamma_2)$, given by (21). To determine the reaction strategy of DM2, we minimize $F(\gamma_2)$ over γ_2. The optimal choice for γ_2 (to be denoted by $\hat{\gamma}_2$) should then satisfy equation (24). Now we do not set $\hat{\gamma}_2 = \gamma_2^t$ in (24) as we did earlier, but instead we solve for $\hat{\gamma}_2$ in terms of γ_2^t and $E[x \mid z_2]$. The resulting expression is

$$\begin{aligned}\hat{\gamma}_2(\eta_2) = {} & [D_{22} - D_{21}\tilde{S} - \tilde{S}^T D_{21}^T + \tilde{S}^T D_{23}\tilde{S}]^{-1} \\ & \times [(\tilde{S}^T(C_{21} - D_{23}D_{11}^{-1}C_{11}) + C_{22})E[x \mid z_2] \\ & + D_{21}D_{11}^{-1}C_{11}E[x \mid z_2] + (D_{21} - \tilde{S}^T D_{23})(D_{11}^{-1}D_{12} - \tilde{S})\gamma_2^t(\eta_2)], \qquad (30)\end{aligned}$$

where $\tilde{S}$ is related to S through (25). In (30) let us replace $\tilde{S}$ by $\alpha\hat{S}$, where $\hat{S}$ is an $(r_1 \times r_2)$-dimensional matrix with the property $\hat{S}^T D_{23}\hat{S} > 0$,† and $\alpha > 0$ is a sufficiently large scalar. Then the dominating term in

$$D_{22} - \alpha D_{21}\hat{S} - \alpha\hat{S}^T D_{21}^T + \alpha^2 \hat{S}^T D_{23}\hat{S}$$

is the last one, and the dominating term in the remaining part of $\hat{\gamma}_2$ is $\alpha^2 \hat{S}^T D_{23}\hat{S}\gamma_2^t(\eta_2)$. Hence, for sufficiently large α, and with $\tilde{S} = \alpha\hat{S}$, $\hat{\gamma}_2(\eta_2)$ behaves like

$$\hat{\gamma}_2(\eta_2) \sim [\alpha^2 \hat{S}^T D_{23}\hat{S}]^{-1}\alpha^2 \hat{S}^T D_{23}\hat{S}\gamma_2^t(\eta_2) = \gamma_2^t(\eta_2),$$

which is the team solution of DM2. Thus, we have shown that if DM1 adopts the strategy

$$\begin{aligned}\gamma_1^\alpha(\eta_1) = {} & -\alpha\hat{S}u_2 - D_{11}^{-1}C_{11}[Kz_1 + (I - KH_1)\Sigma_2 z_2] \\ & + (\alpha\hat{S} - D_{11}^{-1}D_{12})\gamma_2^t(\eta_2), \qquad (31)\end{aligned}$$

where γ_2^t is given by (17b), then the reaction of DM2 is given uniquely by (30) with $\tilde{S}$ replaced by $\alpha\hat{S}$, and this latter sequence converges (under an appropriate norm) to DM2's team strategy γ_2^t. Since (31) is a representation of the team strategy of DM1, for any finite α, it readily follows that by announcing (31) with sufficiently large α, DM1 can come arbitrarily close to the lower bound given in (16); however, it should again be clear from (31) that as $\alpha \to \infty$, the strategy sequence $\{\gamma_1^\alpha(\eta_1)\}$

† It is always possible to find such a matrix $\hat{S}$, since D_{23} is positive definite, and $r_1 \geq r_2$, by hypothesis.

does not approach a well-defined limit. This implies that for $r_1 \geq r_2$, the Stackelberg cost is definitely given by the lower bound of (16); however, it is not always attained, unless the matrix equation (26a) admits a solution.

Remark 3.3. It should be noted that the structural results obtained in this section are not dependent on the Gaussian nature of the statistics of the random variables involved. In particular, expressions (17) could be replaced by the more general versions

$$\gamma_1^t(\eta_1) = -D_{11}^{-1}D_{12}u_2 - D_{11}^{-1}C_{11}E[x \mid z_1, z_2], \tag{32a}$$

$$\gamma_2^t(\eta_2) = -(D_{13} - D_{12}^T D_{11}^{-1} D_{12})^{-1}(C_{12} - D_{12}^T D_{11}^{-1} C_{11})E[x \mid z_2]. \tag{32b}$$

Furthermore, the statement of Theorem 3.2 remains valid with only (29) replaced by

$$\hat{\gamma}_1(\eta_1) = -\tilde{S}^* u_2 - D_{11}^{-1}C_{11}E[x \mid z_1, z_2] + (\tilde{S}^* - D_{11}^{-1}D_{12})\hat{\gamma}_2(\eta_2), \tag{33a}$$

$$\hat{\gamma}_2(\eta_2) = -(D_{13} - D_{12}^T D_{11}^{-1} D_{12})^{-1}(C_{12} - D_{12}^T D_{11}^{-1} C_{11})E[x \mid z_2]. \tag{33b}$$

Remark 3.4. One special version of the decision problem treated in this section is the team problem, in which case $g_1 \equiv g_2$. This requires the parametric restrictions $D_{11} = D_{23}$, $D_{12} = D_{21}^T$, $D_{13} = D_{22}$, $C_{11} = C_{21}$, $C_{12} = C_{22}$. Under these conditions, it can readily be checked that matrices $\tilde{A}$ and $\tilde{B}$ given by (26b) and (26c), respectively, identically vanish, and thus the linear matrix equation (26a) is satisfied for all "representations" $\tilde{S}$, as it should be.

4. Two Applications in Duopoly Markets under Static Information

We now consider applications of the results obtained in Section 2 to two stochastic models of duopoly. In the first model, quantity produced is the decision variable of each firm. Such a model has been considered in Ref. 3 within the context of Nash equilibria. In the second model, however, price is taken as the strategic variable, and properties of such a model have extensively been discussed in Ref. 4, again within the context of Nash equilibria. In what follows, we shall show that the sufficiency conditions derived in Section 2 are satisfied for both models, and thus in each case it is possible to write down the unique hierarchical noncooperative equilibrium strategies of the firms (with Firm 1 taken as the leader in the decision process).

4.1. Stochastic Quantity Model of Duopoly

Using the notation of Ref. 3, wherever appropriate, we first have the linear demand curve

$$p = \alpha - \beta(u_1 + u_2), \tag{34a}$$

where p denotes the price of the commodity, β is a positive constant, and α is a Gaussian random variable with mean $\bar{\alpha}$ and variance σ. The profit functions are given by

$$P_i(\alpha, u_1, u_2) = u_i p - c_i u_i^2, \qquad c_i > 0, i = 1, 2, \tag{34b}$$

and the static information structure provides Firm i (DMi) with the observation

$$z_i = \alpha + w_i, \quad w_i \sim N(0, \lambda_i), \qquad i = 1, 2. \tag{34c}$$

Now, to be consistent with the formulation given in Section 1, we write α as $\alpha - \bar{\alpha} + \bar{\alpha}$, where $\alpha - \bar{\alpha} \triangleq x$ has zero mean. Then, substitution of (34a) into (34b) yields

$$\begin{aligned} P_i(x, u_1, u_2) &= -(\beta + c_i)u_i^2 - \beta u_i u_j + u_i x + u_i \bar{\alpha}, \qquad j \neq i, \\ &\equiv -g_i(x, u_1, u_2), \end{aligned}$$

and thus the weighting terms that define (1a) and (1b) assume the specific values

$$\begin{aligned} &D_{11} = 2(\beta + c_1), \quad D_{12} = D_{21} = \beta, \quad D_{13} = D_{23} = 0, \quad C_{11} = C_{22} = -1, \\ &C_{12} = C_{21} = 0, \quad k_{11} = k_{22} = -\bar{\alpha}, \quad k_{12} = k_{22} = 0, \quad D_{22} = 2(\beta + c_2). \end{aligned}$$

K, defined by (8b), is now given by

$$K = 2(\beta + c_1) - \beta^2/(\beta + c_2) \tag{35}$$

and thus condition (10) becomes

$$0 < 1 - \frac{1}{2}\left(\frac{\beta}{\beta + c_2}\right)\left(\frac{\beta}{\beta + c_1}\right) < 2,$$

which is satisfied for all values of β, c_1, and c_2. To obtain expressions for the unique strategies, we first note that (11a) can be written as

$$\begin{aligned} &\left[2(\beta + c_1) - \left(\frac{\beta^2}{\beta + c_2}\right)\left(\frac{\sigma}{\sigma + \lambda_1}\right)\left(\frac{\sigma}{\sigma + \lambda_2}\right)\right]A \\ &\qquad = \left(\frac{-\beta}{2(\beta + c_1)}\right)\left(\frac{\sigma}{\sigma + \lambda_1}\right)\left(\frac{\sigma}{\sigma + \lambda_2}\right) + \frac{\sigma}{\sigma + \lambda_1}. \end{aligned}$$

Furthermore, q (given by (11c)) becomes

$$q = \frac{\bar{\alpha}}{K}\left(1 - \frac{\beta}{2(\beta + c_2)}\right).$$

Hence, we have the following corollary as a special case of Theorem 2.1.

Corollary 4.1. The stochastic quantity model of duopoly admits a unique hierarchical equilibrium solution under static information structure. The optimal strategies are given by

$$\hat{\gamma}_1(z_1) = \left[2(\beta + c_1) - \left(\frac{\beta^2}{\beta + c_2}\right)\left(\frac{\sigma}{\sigma + \lambda_1}\right)\left(\frac{\sigma}{\sigma + \lambda_2}\right)\right]^{-1}$$
$$\times \left[\frac{\sigma}{\sigma + \lambda_1} - \frac{\beta}{2(\beta + c_2)}\left(\frac{\sigma}{\sigma + \lambda_1}\right)\left(\frac{\sigma}{\sigma + \lambda_2}\right)\right](z_1 - \bar{\alpha})$$
$$+ \left(1 - \frac{\beta}{2(\beta + c_2)}\right)\frac{\bar{\alpha}}{K}, \tag{36a}$$

$$\hat{\gamma}_2(z_2) = \frac{1}{2(\beta + c_2)}\left[\frac{\sigma}{\sigma + \lambda_2}(z_2 - \bar{\alpha}) + \bar{\alpha}\right] - \frac{\beta}{2(\beta + c_2)} E[\hat{\gamma}_1(z_1) \mid z_2], \tag{36b}$$

where K is defined by (35).

4.2. Stochastic Price Model of Duopoly

Considering a possibly nonsymmetric market in which price is the decision variable of each firm, the demand for the product of Firm i is modeled by (see Ref. 4 or Ref. 7)

$$d_i = w_i\beta[V - u_i - \xi(1 - w_i)(u_i - u_j)], \qquad j \neq i, \tag{37a}$$

where u_i denotes the price charged by Firm i, ξ and β are positive constants, V is a Gaussian random variable with mean $\bar{V}$ and variance σ, and w_1, w_2 are weighting factors reflecting the nonsymmetry in the market ($w_1 + w_2 = 1$). The profit functions are given by

$$\pi_i(V, u_1, u_2) = (u_i - c_i)\, d_i - K_i, \tag{37b}$$

where c_i denotes the unit average cost of Firm i and K_i denotes the fixed cost of Firm i. Finally, the static information structure allows Firm i to have access to

$$z_i = V + w_i, \quad w_i \sim N(0, \lambda_i), \qquad i = 1, 2. \tag{37c}$$

Now, substitution of (37a) into (37b) yields

$$\begin{aligned} \pi_i(x, u_1, u_2) &= w_i\beta[-(1 + \xi w_j)u_i^2 + \xi w_j u_i u_j + u_i x \\ &\quad + u_i(c_i(1 + \xi w_j) + \bar{V}) - u_j c_i \xi w_j] - K_i \\ &\equiv g_i(x, u_1, u_2) - K_i, \end{aligned}$$

where $x \triangleq V - \bar{V}$ is a Gaussian random variable with mean zero and variance σ. Hence, for this special version, the weighting terms that define

(1a) and (1b) assume the specific values

$$D_{11} = 2w_1\beta(1 + \xi w_2), \qquad D_{12} = -\xi w_2 w_1 \beta,$$
$$D_{22} = 2w_2\beta(1 + \xi w_1), \qquad D_{21} = -\xi w_1 w_2 \beta,$$
$$C_{11} = -w_1\beta, \quad C_{22} = -w_2\beta, \quad D_{13} = D_{23} = 0, \quad C_{12} = C_{21} = 0,$$
$$k_{11} = (-c_1(1 + \xi w_2) - \bar{V})w_1\beta, \qquad k_{12} = c_1\xi w_2 w_1 \beta,$$
$$k_{22} = -w_2\beta(c_2(1 + \xi w_1) + \bar{V}), \qquad k_{21} = c_2 \xi w_1 w_2 \beta.$$

K is then given by

$$K = 2w_1\beta(1 + \xi w_2) - \frac{\xi^2 w_1^2 w_2 \beta}{1 + \xi w_1}, \tag{38}$$

and thus condition (10) becomes

$$0 < 1 - \frac{1}{2}\left(\frac{\xi w_1}{1 + \xi w_1}\right)\left(\frac{\xi w_2}{1 + \xi w_2}\right) < 2.$$

which is satisfied for all values of ξ, w_1, and w_2. Again, to obtain the relevant unique Stackelberg strategies, we rewrite (11a) and (11c) respectively as follows:

$$\left[2(1 + \xi w_2) - \left(\frac{\xi^2 w_1 w_2}{1 + \xi w_1}\right)\left(\frac{\sigma}{\sigma + \lambda_1}\right)\left(\frac{\sigma}{\sigma + \lambda_2}\right)\right]A$$
$$= \left(\frac{\xi w_2}{2(1 + \xi w_1)}\right)\left(\frac{\sigma}{\sigma + \lambda_1}\right)\left(\frac{\sigma}{\sigma + \lambda_2}\right) + \frac{\sigma}{\sigma + \lambda_1} \tag{39a}$$

and

$$q = \frac{w_1\beta}{K}\left\{\frac{\xi w_2}{2(1 + \xi w_1)}[c_2(1 + \xi w_1) + \bar{V}] - \frac{\xi^2 w_1 w_2 c_1}{2(1 + \xi w_1)} + [c_1(1 + \xi w_2) + \bar{V}]\right\}, \tag{39b}$$

where K is defined by (38). The solution can thus be given as follows as a special case of Theorem 2.1.

Corollary 4.2. The stochastic price model of duopoly admits a unique hierarchical equilibrium solution under static information structure. The optimal strategies are given by

$$\hat{\gamma}_1(z_1) = \left[2(1 + \xi w_2) - \left(\frac{\xi^2 w_1 w_2}{1 + \xi w_1}\right)\left(\frac{\sigma}{\sigma + \lambda_1}\right)\left(\frac{\sigma}{\sigma + \lambda_2}\right)\right]^{-1}$$
$$\times \left(\frac{\sigma}{\sigma + \lambda_1}\right)\left[1 + \left(\frac{\xi w_2}{2(1 + \xi w_1)}\right)\left(\frac{\sigma}{\sigma + \lambda_2}\right)\right](z_1 - \bar{V})q, \tag{40a}$$

$$\hat{\gamma}_2(z_2) = \frac{1}{2(1 + \xi w_1)}\left[\frac{\sigma}{\sigma + \lambda_2}(z_2 - \bar{V}) + \bar{V}\right] + \frac{c_2}{2}$$
$$+ \frac{\xi w_1}{2(1 + \xi w_1)} E[\hat{\gamma}_1(z_1) \mid z_2], \tag{40b}$$

where q is defined by (39b).

5. Concluding Remarks

In this paper, we have proven existence and uniqueness of linear hierarchical noncooperative (Stackelberg) equilibrium strategies for two-person quadratic Gaussian decision problems, and under both static and dynamic information patterns. The dynamic information pattern adopted is of the one-step-delay information sharing type, which allows the leader to have access to both the observation and decision value (but not strategy) of the other decision maker. The ideas introduced and techniques utilized in Section 3 could be effectively used in obtaining the hierarchical noncooperative equilibrium solutions of multistage linear quadratic Gaussian decision problems under the one-step-delay information sharing pattern.

The results obtained in Section 2 under static information structure are then applied to stochastic quantity and price models of duopoly in Section 4. It has been shown earlier that there are certain important structural differences between the two models of duopoly as far as the informational properties of Nash equilibria are concerned (Refs. 3, 4). It would be interesting to investigate whether analogous structural differences exist between the two models in the case of Stackelberg equilibria.

Yet another class of problems that require further investigation is the derivation of Stackelberg equilibria when the dynamic information pattern allows DM1 to have access only to the value of the decision variable of DM2, but not to his observation. Expressions for linear Stackelberg equilibria could be obtained in this case, but whether the linear optimal decision rules are globally optimal is not known at this stage. For an application of this kind of dynamic information within the context of Nash equilibria, the reader is referred to Ref. 6.

References

1. BAŞAR, T., Decentralized multicriteria optimization of linear stochastic systems, *IEEE Transactions on Automatic Control*, Vol. 23, pp. 233–243, 1978.
2. BAŞAR, T., Noncooperative equilibrium solutions in two-person hierarchical decision problems, Marmara Research Institute, internal report, 1978.
3. BAŞAR, T. and HO, Y. C., Informational properties of the Nash solutions of two stochastic nonzero-sum games, *Journal of Economic Theory*, Vol. 7, pp. 370–387, 1974.
4. BAŞAR, T. and YILDIZBAYRAK, N., Noncooperative equilibria and informational properties of a stochastic price model of oligopoly, Marmara Research Institute, internal report, 1978.
5. HO, Y. C., BLAU, I., and BAŞAR, T., A tale of four information structures, in *Lecture Notes in Mathematical Economics and Systems Theory*, No. 107. Edited by A. Bensoussan and J. Lions, Springer Verlag, New York, 1975.

6. Ho, Y. C. and Kastner, M. P., Market signaling: An example of a two-person decision problem with a dynamic information structure, *IEEE Transactions on Automatic Control*, Vol. 23, pp. 350–361, 1978.
7. Levitan, R. E. and Shubik, M., Noncooperative equilibria and strategy spaces in an oligopolistic market, *Differential Games and Related Topics*. Edited by H. W. Kuhn and G. P. Szego, North Holland, Amsterdam, 1971.
8. Witsenhausen, H., Separation of estimation and control for discrete time systems, *Proceedings of IEEE*, Vol. 59, pp. 1557–1566, 1971.

13

A Game-Theoretic Approach to Bureaucratic Behavior

DAGOBERT L. BRITO AND MICHAEL D. INTRILIGATOR

1. Introduction

Bureaucracy is one of the most pervasive forms of organization for modern society. Such organizations, characterized by large size, a hierarchical structure, and few if any concrete measures of performance, exist in government, industry, universities, nonprofit organizations, and other sectors of modern society.† In the U.S. federal government there are, in fact, many "layers" of bureaucracy. Working from the inner to the outer layers, any individual office in an agency is a bureau, as are the agencies themselves and the cabinet department of which they are part. In industry the different divisions or branches of a corporation can be considered bureaus. In universities the departments can be treated as bureaus.

The purpose of this paper is to develop a formal analysis of one critical aspect of bureaucratic behavior, the allocation of resources among different bureaus by means of budgets. Previous work by Niskanen has suggested that bureaus attempt to maximize their budgetary allocations, which provide resources for employment and influence and are considered measures of the absolute and relative importance of the bureaus.‡

† The classic definition and analysis of bureaus is that of Weber. See Downs (Ref. 1) and Mouzelis (Ref. 2) for discussions and extensions of Weber's ideas.

‡ See Niskanen (Refs. 3, 4). For example: "A complete statement of the central motivational assumption of this theory is thus: bureaucrats maximize the total budget of their bureau during their tenure, subject to the constraint that the budget must be equal to or larger than the minimum total costs of supplying the output expected by the sponsor" (Niskanen, Ref. 4, p. 27).

DAGOBERT L. BRITO • Department of Economics, Tulane University, New Orleans, Louisiana. MICHAEL D. INTRILIGATOR • Department of Economics, University of California, Los Angeles, California.

This paper extends the Niskanen analysis by considering the interaction among several bureaus and their relations with allocative and review authorities using concepts of game theory. The next section presents the three classes of agents involved in bureaucratic budgetary allocations and the process of allocation. Section 3 then sets out a description of the behavioral objectives of each of the agents. Section 4 presents the allocative process, and Section 5 summarizes conclusions.

2. The Three Classes of Agents and the Budgetary Process

Three classes of agents are assumed to be involved in the allocative process in a bureaucracy. First, there are the bureaus themselves, and in the theory to be developed it will be assumed that there are two bureaus competing for budgetary allocations.† Second, there is the allocative authority, which is responsible for setting budgets for the two bureaus. Third, there is the review authority, to whom the bureaus can appeal budgetary decisions made by the allocative authority. The bureaus act strategically, but the allocative authority and review authority are assumed to act nonstrategically, resulting in a two-person game.

The three classes of agents can be illustrated using the previously introduced examples of bureaucratic organizations. In the "layers" of bureaucracy in the U.S. federal government different offices in an agency are bureaus, where the head of the agency is the allocative authority and the department is the review authority. At the next higher level the agencies themselves are bureaus, where the allocative authority is the department and the review authority is the Office of Management and Budget (OMB). Among departments, such as the Department of Defense and the Department of Health, Education, and Welfare, each regarded as a bureau, the allocative authority is OMB and the review authority is Congress. In industry the different divisions of a corporation can be treated as bureaus, where the allocative authority is the central management (chief executive officer) and the review authority is the board of directors or the stockholders. Finally, in a university the departments can be treated as bureaus, with the dean as the allocative authority and the provost or chancellor as the review authority.

The process of budgetary allocation occurs at fixed budget review points, e.g., annually. The budgets of the bureaus are initially set by the allocative authority, which is responsible for allocating a fixed aggregate budget. If the allocation proposed by the allocative authority is not

† Restricting attention to two bureaus avoids the problem of coalitions forming among two or more bureaus, e.g., two bureaus coordinating their actions to take advantage of a third bureau.

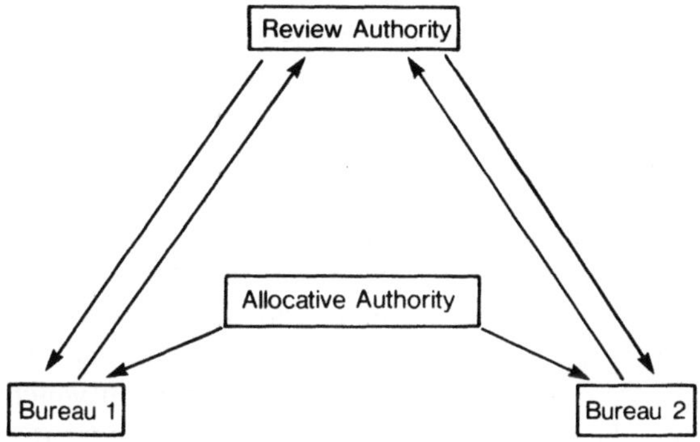

Figure 1. Relationship among the three classes of agents.

accepted by the bureaus, then the budget is allocated by the review authority. For example, if one or both of the bureaus does not accept the budget assigned to it by the allocative authority, it can lobby the review authority.† If the review authority is brought into the process, however, it may change the aggregate budget for both bureaus as well as modify the individual budgets. In particular, the review authority can review all budgets, individually and collectively, and, as a result, reduce the total budget, leading possibly to lower budgets for both bureaus. The bureaus thus must choose between accepting a given budget from the allocative authority or, alternatively, appealing to the review authority, which may augment its budget but may also lower all budgets.

From the examples given there are two types of review authorities. Either the review authority can be a single agent (e.g., a federal department, a provost or chancellor) or it can be a set of agents (e.g., Congress, stockholders in a corporation). In the theory to be developed it will be assumed that the review authority is a set of agents, but it will be shown that this set of agents can be replaced by a single agent, representing the median voter in the set of agents.

Figure 1 shows the relationships among the three classes of agents. The allocative authority allocates a fixed aggregate budget, determined by

† For discussions of how this process works in practice see Wildavsky (Ref. 5), Tullock (Ref. 6), and Gawthrop (Ref. 7). For example, quoting Gawthrop, p. 169, "committees of Congress are very much like interest groups in that they too have valuable resources at their disposal that can be effectively utilized to supplement the support received by the executive official. A bureau chief or even an Assistant Secretary will, in all probability, not be able to alter a budget decision made within the executive branch by the Bureau of the Budget in connection with the individual's own administrative operations. Either one of these individuals, on the other hand, may be quite capable of having the deleted funds restored by simply appealing—indirectly of course—to 'his' congressional subgroup."

information contained in last year's budget, between the two bureaus. The bureaus can bring the review authority into the process, however, in which case both the individual budgets and the aggregate budget can be changed. The figure can be applied to the previous examples of federal departments, such as Departments of Defense and Health, Education and Welfare, where the Office of Management and Budget is the allocative authority and Congress is the review authority. It can also be applied to a state bureaucracy, where the bureaus can be highways and health, the allocative authority is the state budget authority, and the review authority is the governor. Another example is a corporation in which the bureaus are the divisions, the allocative authority is the chief executive officer, and the review authority is the board of directors. Other examples include the university and muncipal government.

3. Behavioral Objectives of the Agents

Each of the three agents involved in the allocative process in a bureaucracy is assumed to solve a maximization problem summarizing its objectives and the constraints it faces. The solution to each of these problems indicates the behavior of the relevant agent.

Each of the bureaus is assumed to maximize a welfare functional subject to budget and technical constraints. Bureau i $(i = 1, 2)$ is assumed to maximize a welfare functional at time τ of the form

$$B_i = \int_\tau^T b_i(x_i, v_i, w_i)\, dt, \tag{1}$$

representing accumulated utility from the present τ to the time horizon T. The time horizon T is the time when the bureau is scheduled for a regular review by the review authority. The instantaneous utility function at time t, given as $b_i(x_i(t), v_i(t), w_i(t))$ depends on x_i, the output of public good i produced by the bureau; on v_i, its spending on the first input, and on w_i, its spending on the second input. The two inputs might, for example, be spending on labor and capital inputs used to produce the public good. Alternatively, they might be spending on different programs used to provide the public good, such as Air Force spending on missiles and bombers used to provide national security.† The utility function $b_i(\cdot)$ is assumed concave, and with positive first-order partial derivatives in all arguments.

† Yet another possible interpretation might be one in which v_i is the spending on inputs needed to produce the public good while w_i is the level of spending on perquisites for bureaucrats.

In period k, extending from time k to time $k + 1$, bureau i solves the problem

$$\max_{\{v_i(t),\, w_i(t)\}} \int_k^{k+1} b_i(x_i, v_i, w_i)\, dt, \tag{2}$$

that is, the problem of maximizing the component of B_i in period k by choice of $v_i(t)$ and $w_i(t)$. This maximization problem is solved subject to two constraints. The first is the budget constraint

$$y_i(k) = \int_k^{k+1} [v_i(t) + w_i(t)]\, dt, \tag{3}$$

relating the budget of bureau i in period k, given as $y_i(k)$, to its spending on the two inputs in this period. The second constraint is the production function†

$$x_i = F_i(v_i, w_i), \tag{4}$$

relating the output of the public good to the spending on the two inputs. The function $F(\cdot)$ is assumed concave in both arguments, with positive first-order partial derivatives.

With this formulation of the behavioral objectives and constraints for the bureaus, the welfare of each bureau, to be maximized, depends not only on the time path of its budget but also on the time path of the level of the public goods that it produces and on the means by which these public goods are provided. Previous work by Niskanen is thus generalized in two ways—first, by allowing welfare to depend not only on the instantaneous budget but also on future budgets and, second, by allowing utility to depend also on the level of the public good provided and on the division of the budget between different inputs used in the production of the public good.

The *tax price* of the ith public good at the end of period k will be defined as the ratio of the budget of bureau i to the cumulative amount of the public good provided in this period:

$$p_i(k) = y_i(k) \Big/ \int_k^{k+1} x_i(t)\, dt. \tag{5}$$

It will be assumed for simplicity that no other agent is able to observe the expenditure behavior of a bureau. The only observable variables are thus assumed to be the supply of the public good and budgets, or, equivalently, the budgets of the bureaus and the tax prices in (5). Thus letting $m(k)$ be the vector

$$m(k) = [y_1(k), y_2(k), p_1(k), p_2(k)] \in M(k), \tag{6}$$

† With the interpretation of the last footnote, in which w_i is spending on perquisites (4) would become $x_i = F_i(v_i)$.

where $M(k)$ is the set of all possible combinations of budgets and tax prices in period k, the *information set* of all agents at time k is given as

$$N(k) = \mathop{\times}_{r=0}^{k} M(r), \tag{7}$$

that is, as the set of all past budgets and all past tax prices, starting at historical initial time.

The solution to the problem of bureau i in the kth period, namely

$$B_i(k) = \max_{\{v_i(t), w_i(t)\}} \int_k^{k+1} b_i(x_i, v_i, w_i)\, dt \tag{8}$$

subject to

$$y_i(k) = \int_k^{k+1} (v_i + w_i)\, dt, \tag{9}$$

$$y_i(k) = p_i(k) \int_k^{k+1} F_i(v_i, w_i)\, dt = p_i(k) \int_k^{k+1} x_i(t)\, dt, \tag{10}$$

defines a unique path for $v_i(t)$, $w_i(t)$, and $x_i(t)$. Thus the welfare of bureau i in period k depends on the tax price and budget of the bureau in this period, i.e.,

$$B_i(k) = B_i[p_i(k), y_i(k)]. \tag{11}$$

The allocative authority is assumed to maximize a welfare functional at time τ of the form

$$A = \int_\tau^{T_A} a(x_1, x_2)\, dt, \tag{12}$$

which is, again, the accumulated utility over the horizon starting at τ and ending at T_A. The instantaneous utility function at time t, given as $a(x_1(t), x_2(t))$, depends on the public good produced by each of the two bureaus, x_1 and x_2. This utility function for the allocative authority is also assumed concave, with positive first-order partial derivatives. In period k the allocative authority solves the problem of finding

$$A(k) = \max_{\{y_1(t), y_2(t)\}} \int_k^{k+1} a(x_1, x_2)\, dt \tag{13}$$

subject to the budget constraint requiring that total spending be based on public goods provided and the past levels of tax prices

$$Y(k) = p_1(k-1)x_1 + p_2(k-1)x_2 = y_1(k) + y_2(k), \tag{14}$$

where x_1 and x_2 are the steady-state amounts of the two public goods.

The solution to this problem is a function of the current tax prices and the current budgets

$$A(k) = A[p_1(k), p_2(k), y_1(k), y_2(k)] \tag{15}$$

since, given the current tax prices and budgets, the outputs of the two public goods can be determined and used to compute $A(k)$ in (13).

The review authority is assumed to be a set of individuals, such as Congress or stockholders of a corporation, where the case of a single entity as the review authority can be treated as a special case. Assuming a continuum of individuals characterized by the parameter n, where n is assumed to be distributed uniformly in the unit interval [0, 1], at time τ agent n maximizes the welfare functional

$$R(n) = \int_{\tau}^{T_R} r[nx_1, (1-n)x_2, nz]\, dt. \tag{16}$$

Here the instantaneous utility function $r(\cdot)$ depends on the supply of the two types of public goods x_1, x_2 and also on z, a scalar measure of all private goods. This utility function is assumed concave, with positive first-order partial derivatives and with the cross partials r_{x_1z}, r_{x_2z} assumed to be zero. The parameter n characterizes the way that review agents differ in their preferences for either the two public goods or the private good. In period k agent n of the review authority solves the problem

$$R(k, n) = \max_{\{y_1(t), y_2(t)\}} \int_k^{k+1} r[nx_1, (1-n)x_2, nz]\, dt \tag{17}$$

subject to the overall budget constraint

$$p_1(k-1)x_1 + p_2(k-1)x_2 + z = 1, \tag{18}$$

where y_i has been replaced by $p_i(k-1)x_i$. This overall budget constraint carries forward the tax prices of the previous period, weights them by the outputs of the two public goods, and adds the level of private goods, which acts as numeraire and thus has a price of unity. The total is the total available budget, normalized as unity. The budget constraint for the review authority (18) can be contrasted to that for the allocative authority (14). Both use past tax prices and levels of public goods to determine available budgets, but the review authority has the added option of expanding or contracting the total budget for both bureaus by contracting or expanding the level of private goods z. The solution to the problem of agent n in the review authority in period k can be written as

$$R(k, n) = R[p_1(k), p_2(k), y_1(k), y_2(k), n], \tag{19}$$

since, as before, the current tax prices and budgets permit determination of the outputs of the public goods, which, together with the budget constraint, determine the private good and thus yield sufficient information for computation of the integral in (17).

If resources are allocated by the review authority in budget period k, then the review agents vote costlessly on proposals pairwise until one is found that cannot be dominated. This process is costless, and each agent assumes the tax price of the ith bureau to be $p_i(k-1)$, as in (18). Assuming nonstrategic voting by all members of the review authority, an important property of the allocative process of the review authority is that it is determined by the median voter in the class of review agents. Thus the solution to the allocative process by the review authority is that it reduces to that of a single-agent authority where the utility function is

$$r[\tfrac{1}{2}x_1, \tfrac{1}{2}x_2, \tfrac{1}{2}z]. \tag{20}$$

Any other proposal will be dominated by this outcome. Thus the set of agents in the review authority can be replaced by a single agent with utility function at the median level of all agents.

To prove this median voter result, note first that the nonstrategic voting assumption leads to consideration of steady states, in which case the instantaneous utility function $r[nx_1, (1-n)x_2, z]$ is maximized subject to the (normalized) budget

$$p_1(k-1)x_1 + p_2(k-1)x_2 + z = 1. \tag{21}$$

Letting $\bar{x}_1$, $\bar{x}_2$, $\bar{z}$ be the solution for $m > n$, where

$$\bar{r} = r[n\bar{x}_1, (1-n)\bar{x}_2, n\bar{z}], \tag{22}$$

and differentiating with respect to m gives

$$\frac{d\bar{r}}{dm} = r_1 n \frac{d\bar{x}_1}{dm} + r_2(1-n)\frac{d\bar{x}_2}{dm} + r_3 n \frac{d\bar{z}}{dm}. \tag{23}$$

From (21), however

$$\frac{d\bar{x}_2}{dm} = -\frac{p_1}{p_2}\frac{dx_1}{dm} - \frac{1}{p_2}\frac{d\bar{z}}{dm}, \tag{24}$$

so

$$\frac{d\bar{r}}{dm} = \left[nr_1 - \frac{p_1}{p_2}(1-n)r_2\right]\frac{d\bar{x}_1}{dm} + \left[nr_3 - \frac{1}{p_2}(1-n)r_2\right]\frac{d\bar{z}}{dm}. \tag{25}$$

But $d\bar{x}_1/dm$ and $d\bar{z}/dm$ are positive, so $d\bar{r}/dm$ will be negative if both square bracket terms in (25) are negative, which follows from the first-order (Kuhn–Tucker) conditions of maximizing r subject to (21). Thus if $m > n$, then

$$\frac{d\bar{r}}{dm} < 0, \tag{26}$$

showing that any move away from any arbitrary n voter by making m larger makes that voter worse off. The outcome is therefore at the median voter, as in (20).

The median voter result for the review authority will give rise to a set of budgetary allocations for the bureaus, called the "threat budgets," summarized by the threat vector in period k

$$y^t(k) = [y_1^t(k), y_2^t(k)]. \tag{27}$$

This is the outcome if the budgetary decision is made by the review authority. Since x_1, x_2, and z are all assumed to be normal goods, it follows from the Slutsky equation that if x_i^t and z^t are the outputs of the public goods and the private good at the threat point, then

$$\frac{\partial x_i^t}{\partial p_i} < 0, \qquad i = 1, 2, \tag{28}$$

$$\frac{\partial x_i^t}{\partial p_j} > 0, \qquad i = 1, 2, \quad i \neq j, \tag{29}$$

$$\frac{\partial z^t}{\partial p_i} > 0, \qquad i = 1, 2. \tag{30}$$

By the budget constraint, however, the last inequality implies that

$$\frac{\partial y_i^t}{\partial p_i} < 0, \qquad i = 1, 2. \tag{31}$$

Thus an increase in the tax price of any public good decreases the threat output of that public good and the threat budget of the bureau producing that public good, and it increases the threat output of the other public good and the threat output of the private good.

4. The Allocative Process

The allocative process begins with the allocative authority, which in period k solves the maximizing problem in (13) subject to the budget constraint in (14). The solution for the budgets allocated to the bureaus in period k is

$$y^a(k) = [y_1^a(k), y_2^a(k)], \tag{32}$$

where, from (14),

$$y_1^a(k) + y_2^a(k) = Y(k). \tag{33}$$

Each of the bureaus will, however, use the information set $N(k)$ in

(7) to project an expected budget for itself, leaving the residual to the other bureau.† Thus bureau 1 will project a budget for itself of $\hat{y}_1^1(k)$, entailing the budgetary allocation

$$\hat{y}^1(k) = [\hat{y}_1^1(k), Y(k) - \hat{y}_1^1(k)]. \tag{34}$$

Similarly bureau 2 will propose the allocation

$$\hat{y}^2(k) = [Y(k) - \hat{y}_2^2(k), \hat{y}_2^2(k)]. \tag{35}$$

In general it can be expected that the budget each bureau projects for itself, based on past levels of budgets and tax prices, will exceed the budget provided by the allocative authority

$$\hat{y}_1^1(k) > y_1^a(k), \qquad \hat{y}_2^2(k) > y_2^a(k). \tag{36}$$

Each bureau then determines its (dis)satisfaction with the budget provided by the allocative authority and, on this basis, decides the extent to which it will lobby for a review by the review authority. Using mixed strategies, bureau 1 chooses a probability $\alpha_1^1(k)$ that it will engage in lobbying for a review, and bureau 2 chooses a probability $\alpha_2^2(k)$ that it will engage in lobbying. Thus bureau 1 chooses both a budgetary allocation $\hat{y}^1(k)$, as in (34) and a probability vector

$$\alpha^1(k) = [\alpha_1^1(k), 1 - \alpha_1^1(k)], \tag{37}$$

indicating its probability of lobbying for a review. Similarly bureau 2 chooses both $\hat{y}^2(k)$, as in (35), and the probability vector

$$\alpha^2(k) = [1 - \alpha_2^2(k), \alpha_2^2(k)]. \tag{38}$$

If one bureau lobbies and the other does not, then the lobbying bureau will obtain its desired budget, with the other receiving the residual. If neither lobbies, then they both must accept the budgets of the allocative authority. If both lobby, however, then the review authority will choose the outcome as the threat budgets of (27), which can be assumed to be less for each bureau than the budget proposed for it by the allocative authority.‡

The bargaining process in terms of the probabilities of lobbying then has the structure of a two-person nonzero-sum game, as summarized in

† For a discussion of one type of projection that could be used and its implications see Brito and Intriligator (Ref. 8). For a discussion of how federal departments set budget requests in practice see Gawthrop (Ref. 7).

‡ If the threat budget exceeds the budget of the allocative authority, then the bureau would go directly to the review authority, circumventing the allocative authority.

	α_1^1	$1 - \alpha_1^1$
$1 - \alpha_2^2$	$\hat{y}_1(k)$ (Budgets proposed by bureau 1)	$y^a(k)$ (Budgets of allocative authority)
α_2^2	$y^t(k)$ (Budgets of review authority)	$\hat{y}_2(k)$ (Budgets proposed by bureau 2)

Figure 2. Outcomes of the bargaining process.

Fig. 2. The outcomes are

$$\begin{aligned} &\hat{y}^1(k), && \text{with probability} \quad \alpha_1^1(1 - \alpha_2^2), \\ &\hat{y}^2(k), && \text{with probability} \quad (1 - \alpha_1^1)\alpha_2^2, \\ &\hat{y}^a(k), && \text{with probability} \quad (1 - \alpha_1^1)(1 - \alpha_2^2), \\ &\hat{y}^t(k), && \text{with probability} \quad \alpha_1^1\alpha_2^2. \end{aligned} \tag{39}$$

It is, in fact, a game of brinksmanship, where each bureau may be tempted to lobby to increase its budget by choosing large values of α_1^1 or α_2^2, but if either or both lobbies "too much" the outcome can be disastrous, each receiving the threat budgets.† If the bureaus are cautious and do not lobby significantly, then they will receive the allocative authority budgets.

This process has a structure that is similar to one developed in connection with work on the Nash bargaining problem.‡ It is illustrated in Fig. 3, showing the budgets for each bureau y_1 and y_2, the budgets proposed by the allocative authority y^a (the status quo or initial allocation in the Nash bargaining problem), the budget proposed by each bureau $\hat{y}^1$ and $\hat{y}^2$, and the threat budgets of the review authority y^t. From previous results on the Nash bargaining problem it follows that the outcome of the bargaining process is determined by an allocation function

$$y^f = \phi(y^a, y^t), \tag{40}$$

where y^f is the budgetary allocation conditional on threats not being

† For discussion of this game in another context see Brito and Intriligator (Ref. 9).

‡ See Brito, Buoncristiani, and Intriligator (Ref. 10). The axioms used in this new approach to the Nash bargaining problem are the Nash axioms of feasibility, individual rationality, Pareto optimality, independence of monotonic linear transformations of utility, and independence of irrelevant alternatives. The Nash axiom of symmetry, however, is dropped and an additional axiom of differentiability (of the allocation function) is assumed.

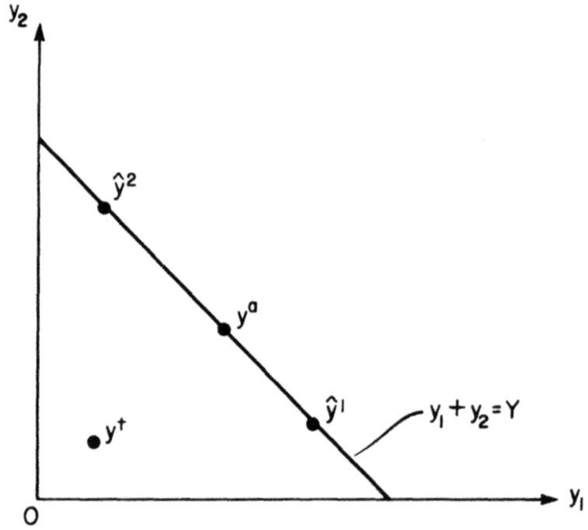

Figure 3. The various budget allocations.

implemented. In the neighborhood of a solution the final allocation varies with the threat allocation

$$\frac{\partial y_i^f}{\partial y_i^t} \geq 0 \quad \text{and} \quad \frac{\partial y_i^f}{\partial y_j^t} \leq 0, \tag{41}$$

and it satisfies the Slutsky-type equation

$$\frac{\partial y_i^f}{\partial y_i^t} + \frac{\partial y_i^f}{\partial y_i^a} = 1 + \frac{\partial y_i^f}{\partial y_j^t}, \tag{42}$$

where the first term on the left is like a substitution effect and the second is like an income effect. Combining (41) and (42) yields

$$\frac{\partial y_i^f}{\partial y_i^a} \leq 1; \tag{43}$$

however, while this derivative is bounded above by unity, it can be negative. Increasing the threat allocation in one bureau's favor therefore never decreases its final allocation, but increasing the allocative authority allocation in one bureau's favor may decrease its final allocation. If the final allocation does increase in this case there can be no overreaction, the change being less than that in the allocative authority allocation. It should also be noted that changes in the projected budgets of the bureaus, $\hat{y}^1$ and $\hat{y}^2$, will influence the final budgetary allocation only indirectly—to the extent that they change the allocations of either the allocative authority or the review authority.

Another previous result for the Nash bargaining problem that applies to this problem of budgeting is that the process converges over time to the allocation of the allocative authority.† The linkage from past prices to current allocation ensures this convergence. Thus the bureaucracy is eventually controllable by the allocative authority, so that in the long run the review authority has no influence other than its influence on the allocative authority.

These results can be applied to the effects on the budgets of a bureaucracy when there is a change in either the allocative authority or the review authority. In the application to the U.S. federal government, for example, election of a new president will generally lead to a change in the allocative authority (the Office of Management and Budget) but need not entail a change in the review authority (Congress). Election of a new Congress, for example, the midterm Congressional elections, can conversely result in a new review authority without a change in the allocative authority.

If there is a change in the allocative authority, such as the election of a new president, there is a change in the budgets proposed by the allocative authority y^a, without a change in the threat budgets of the review authority. With no change in the threat budget the short-run effect of this change is anticipated to be minor. In the long run, however, the change is expected to be large, given the convergence of the budgeting process to the allocative authority's proposed budgets. These results are consistent with the frequent observation that it takes time for a new government to gain control over the bureaucracy.

The other change is one in the review authority, such as the election of a new Congress. In this case there is a change in the threat point, reflecting the different preferences of the new Congress (in particular, the new median voter in Congress). The result of such a change in the threat point will generally be a change in the final outcome budgets in the short run. In the long run, however, the effect of the new review authority disappears because of the offsetting influence of the allocative authority. The allocative authority can, by changing its allocation, counter whatever changes occur in the threat point. Thus this case is just the opposite of the previous one, there being only a transitory effect of a change in the review authority. In particular a new Congress must recognize that its budgetary impacts are primarily short-run ones.

5. Conclusions

This paper has explored the application of game-theoretic approaches to the determination of budgets for bureaucratic agencies. In

† This asymptotic result assumes "structural stability" of the allocation (i.e., no learning phenomena or other dynamic shifts in the allocation function) and "normality" of the allocation function (i.e., positivity of the partial derivatives in (43)), as discussed in Brito, Buoncristiani, and Intriligator (Ref. 10).

this approach budgets are set by an allocative authority, but bureaus can appeal to a review authority by lobbying efforts. A high lobbying effort on the part of both bureaus would generate an audit, in which case the review authority would set the budgets at low threat levels. Final budgets are determined on the basis of both the budgets set by the allocative authority and those constituting the threat of the review authority. While the review authority can have substantial impact in the short run, in the long run the allocative authority determines the budgets.

While this approach can explain some observed bureaucratic behavior, such as the fact that it takes some time for a new allocative authority to have impact, it is believed that the approach can explain other observed aspects of bureaucratic behavior as well. For example, the approach can be used to study the allocation over the budget period to explain why bureaus frequently retain their funding until near the end of the period and then spend most or all of it in the last part of the period.

ACKNOWLEDGMENT

Support of the National Science Foundation in the form of collaborative research grants is gratefully acknowledged.

References

1. DOWNS, A., *Inside Bureaucracy*, Little, Brown and Co., Boston, 1968.
2. MOUZELIS, N. P., *Organization and Bureaucracy*, Aldine Publishing Co., Chicago, 1973.
3. NISKANEN, W. A., Jr., *Bureaucracy and Representative Government*, Aldine–Atherton, Chicago, 1971.
4. NISKANEN, W. A., Jr., *Bureaucracy: Servant or Master?* Institute of Economic Affairs, London, 1973.
5. WILDAVSKY, A., *The Politics of the Budgetary Process*, Little Brown and Co., Boston, 1964.
6. TULLOCK, G., *The Politics of Bureaucracy*, Public Affairs Press, Washington, D.C., 1965.
7. GAWTHROP, L. C., *Bureaucratic Behavior in the Executive Branch*, Free Press, New York, 1969.
8. BRITO, D. L., and INTRILIGATOR, M. D., A fixed point approach to multiagent adaptive control, *Annals of Economic and Social Measurement*, Vol. 6, pp. 137–145, 1977.
9. BRITO, D. L., and INTRILIGATOR, M. D., Strategic weapons and the allocation of international rights, in *Mathematical Systems in International Relations Research*. Edited by J. V. Gillespie and D. A. Zinnes, Praeger Publishers, New York, 1975.
10. BRITO, D. L., BUONCRISTIANI, A. M., and INTRILIGATOR, M. D., A new approach to the Nash bargaining problem, *Econometrica*, Vol. 45, pp. 1163–1172, 1977.

14

Optimal Stabilization Policies for Decentralized Macroeconomic Systems with Conflicting Targets

HAJIME MYOKEN

1. Introduction

The relationships between economic variables in large-scale models are quite complex, and a realistic policy and planning scheme is definitely characterized by an informationally decentralized structure. On the most general level, the state of the system is affected by the actions of several decision makers. The solution to such a problem is possible only under certain decentralization of the policy planning decision making. On the other hand, the traditional policy and planning (Refs. 1, 2) are handled in a centralized way that is set up by the optimal control problem with classical information structure (Ref. 3). In such a scheme, all the information necessary to make an optimal decision is assumed to be available to the central planning agency.

Recent years have witnessed interesting developments of decentralized control methods for large-scale systems, including the stochastic control problem with nonclassical information structure (Ref. 3; See Ref. 4 also). As is seen, for example, in dynamic team theory (Ref. 5) or decentralized control systems (Ref. 6), studies along these lines are emphasized on theoretical and analytical aspects of the optimal linear quadratic Gaussian stochastic control decision problem. However, even

HAJIME MYOKEN • Faculty of Economics, Nagoya City University, Mizuhocho Mizuhoku, Nagoya, Japan.

for small economic systems, this approach has heavy computational requirements which are still difficult to compute and to implement. A relevant example is the literature on stability and stabilizability of decentralized systems which are related to the problem considered in this paper. The studies on the decentralized macroeconomic policy and planning scheme are still in unexplored fields and few in number, except for the related works (Refs. 7–9).

It may be dangerous to regard the stabilization problem of a real decentralized economic system in the same light with the problem of decentralized systems developed in the engineering control literature. To see the stability of decentralized economic system let us now suppose that the ith control agency makes the policy decision specified by

$$x_i(t + 1) = K_i(y_i(t) - y_i^*).$$

Then the desired target level is either achieved or not achieved. In such a case, the system converges to the desired target level and is stable only when there is coordination between the control agencies at the setting of the target level. Thus stabilization policies must be investigated by considering the bias of the equilibrium values and the target values.

The outline of the paper is as follows. Sections 2 and 3 derive the bias of both decentralized policy rules in static and dynamic systems and investigate stability properties and stabilization policies. In addition, it is verified that the biases of a static system and a dynamic system are identical. Section 4 shows how the feedback parameter structured in the decentralized policy rule is determined. The decentralized operations of the well-known Klein's Model I (Ref. 10) for the U.S. economy are numerically illustrated in Section 5.

2. The Bias of Decentralized Policy Rule (1): The Static Case

Consider the following linear, time-invariant model of the static system:

$$y(t) = Bx(t), \tag{1}$$

where $x(t) \in R^n$ and $y(t) \in R^n$ are the vectors of control (instrument) and endogenous variables, respectively, and $\det B \neq 0$. When the policy rule is given as

$$x(t + 1) = K[y(t) - y^*], \tag{2}$$

then it is shown from (1) and (2) that

$$y(t + 1) = BK[y(t) - y^*]. \tag{3}$$

Accordingly,

$$\lim_{t \to \infty} y(t) \neq y^*, \quad \text{for} \quad y^* \neq 0.$$

This implies that the policy rule (2) does not converge to the desired target vector y^*.

Now suppose that the policy rule is given as

$$\Delta x(t) = K[y(t) - y^*], \tag{4}$$

where $\Delta x(t) = x(t + 1) - x(t)$, $\det K \neq 0$. Then the target rule is found from (1) and (4) to be

$$y(t + 1) - y^* = (I + BK)[y(t) - y^*]. \tag{5}$$

As far as

$$|\lambda(I + BK)| < 1 \tag{6}$$

is satisfied, then

$$\lim_{t\to\infty} y(t) = y^*, \qquad \text{for any } y^*.$$

There exists K *satisfying* (6), when the matrix B is regular. In a decentralized system, however, all the information about $y(t)$ is not always available to the decision maker (a government agency, say). In the simplest case where K is restricted to be diagonal, McFadden (Ref. 7) demonstrated that a sufficient condition for the existence of K satisfying (6) is that B has a nested sequence of nonzero principal minors.†

It is assumed that in a real decentralized economic system each control agency is decentralized but some target vairables for each control agency are overlapped; conflicting targets are considered. Let x_1 and x_2 be control variables, and let y_I^* and y_{II}^* be target variables. The decentralized policy rule corresponding to (4) is

$$\begin{pmatrix} \Delta x_1(t) \\ \Delta x_2(t) \end{pmatrix} = \begin{pmatrix} K_1[y(t) - y_I^*] \\ K_2[y(t) - y_{II}^*] \end{pmatrix}, \tag{7}$$

where y_I^* and y_{II}^* are overlapped. Now let the desired target vector be denoted by y^*. Then (7) is given as

$$\Delta x(t) = K[y(t) - y^*] - h, \tag{8}$$

where

$$K = \begin{pmatrix} K_1 \\ K_2 \end{pmatrix}, \qquad h = \begin{pmatrix} K_1(y_I^* - y^*) \\ K_2(y_{II}^* - y^*) \end{pmatrix}.$$

Thus from (1) and (8) we obtain

$$y(t + 1) - y^* = (I + BK)[y(t) - y^*] - Bh. \tag{9}$$

† For the definition regarding this, see McFadden (Ref. 7). Refer to the Fisher–Fuller theorem (Ref. 11) also.

Therefore, there exists K satisfying (6), as far as $\det(BK) \neq 0$. Consequently, we have

$$y(t) - y^* = (I + BK)^t(y_0 - y^*) + (BK)^{-1}[I - (I + BK)^t]Bh. \quad (10)$$

The equilibrium vector, denoted by y^e, is then defined by

$$y^e = \lim_{t\to\infty} y(t) = y^* + K^{-1}h. \quad (11)$$

Next, we consider the bias of the equilibrium values and the target values arising from no coordination in the decision makers of decentralized systems. The bias, denoted by y^{bias}, is defined by

$$y^{\text{bias}} = y^e - y^* = K^{-1}h. \quad (12)$$

Now, letting

$$K^{-1} = (K^1, K^2),$$

the bias is given as

$$y^{\text{bias}} = K^1K_1(y_I^* - y^*) + K^2K_2(y_I^* - y^*).$$

Thus, for the first control agency the bias is

$$y_{\mathrm{I}}^{\text{bias}} = y^e - y_{\mathrm{I}}^* = K^2K_2(y_{\mathrm{II}}^* - y_{\mathrm{I}}^*). \quad (13)$$

From $K_1K^2 = 0$, we have $K_1y_{\mathrm{I}}^{\text{bias}} = 0$. The preceding results imply that for the first control policy the bias is in the null space of the column of feedback parameter K_1.

3. The Bias of Decentralized Control Rule (2): The Dynamic Case

Consider a discrete-time linear dynamic system represented by

$$y(t) = Ay(t - 1) + Bx(t), \quad (14)$$

where $x(t)$, $y(t)$, and B are the same as the ones used in the static case (1). Let the policy rule be given as

$$x(t + 1) = K[y(t) - y^*] + B^{-1}(I - A)y^*. \quad (15)$$

Then we have

$$y(t + 1) - y^* = (A + BK)[y(t) - y^*]. \quad (16)$$

Accordingly, as far as

$$|\lambda(A + BK)| < 1 \quad (17)$$

is satisfied, then

$$\lim_{t \to \infty} y(t) = y^*, \qquad \text{for any } y^*.$$

Also, there exists K satisfying (17), as far as $\det B \neq 0$.

The policy rule (15) has the terms of $B^{-1}(I - A)y^*$ which assume the coordination between the control agencies. Then (15) is not regarded as the policy rule of a decentralized system. We consider therefore the decentralized policy rule shown in (7).

Before proceeding to the discussion, we introduce the shift operator, defined by $z^{\pm i}x(t) = x(t \pm i)$. Using the shift parameter, we obtained the following equation from (7) and (14):

$$y(t + 1) - y^* = [I + z(zI - A)^{-1}BK][y(t) - y^*] - z(zI - A)^{-1}Bh. \quad (18)$$

If the generalized eigenvalue of

$$I + z(zI - A)^{-1}BK \quad (19)$$

is less than unity in the absolute value, then

$$\lim_{t \to \infty} y(t) = y^*, \qquad \text{for any } y^*.$$

The generalized eigenvalues are easily shown to be the roots of

$$\det\{\lambda I - [I + \lambda(\lambda I - A)^{-1}BK]\} = 0, \quad (20)$$

which can be written as

$$\frac{\det\{\lambda^2 I - \lambda(I + A + BK) + A\}}{\det(\lambda I - A)} = 0. \quad (21)$$

When considering the numerator of (21) only, then it is equal to the eigenvalues of

$$M = \begin{pmatrix} 0 & I \\ -A & I + A + BK \end{pmatrix}. \quad (22)$$

As far as $\det(BK) \neq 0$, the matrix M in (22) does not have the eigenvalue of unity. Therefore, the generalized eigenvalues of (19) are identical to the $2n$ eigenvalues of the matrix M. Consequently, if $|\lambda(M)| < 1$, it then is shown from (18) that

$$y(t) - y^* = [I + z(zI - A)^{-1}BK]^t(y - y^*) + [z(zI - A)^{-1}BK]^{-1}[I - \{I + z(zI - A)^{-1}BK\}^t]z(zI - A)^{-1}Bh. \quad (23)$$

From the results obtained before, it is found that the bias of the static system is identical with that of the dynamic system.

4. Determination of Feedback Parameter

In this section I briefly describe how the feedback parameter is determined in practice. This problem is considered only for the first control agency. Let

$$B = (B_1, B_2), \qquad b_2(t) = B_2(t) + x_2(t).$$

Further, instead of (14), a linear model for the first control agency is assumed to be given as

$$y(t) = Ay(t-1) + B_1 x_1(t) + b_2(t). \tag{24}$$

And suppose that a quadratic loss function is specified by

$$\phi_1 = \sum_{t=1}^{T} [y_1(t) - y_1^*(t)]' Q_1(t) [y_1(t) - y_1^*(t)]. \tag{25}$$

The optimal linear feedback control equation is computed using well-known methods as follows:

$$x_1(t) = g_1(t) + G_1(t) y(t-1). \tag{26}$$

The results of (26) correspond to those of the optimal control problem developed by Chow (Ref. 12). In the optimal feedback control equation (26), $g_1(t)$ is the feedforward vector, which is determined by the policy forecasting value b_τ ($\tau = t, t+1, \ldots, T$) taken by the second control agency and the target level in the future y_τ^* ($\tau = t, t+1, \ldots, T$). Also, $G_1(t)$ is the feedback matrix, which is computed in the combination with a matrix Ricatti equation.

5. Optimal Decentralized Stabilization for Klein's Model I

5.1. Klein's Model I

In this section we investigate numerically the problems connected with the decentralized stabilization proposed in the preceding sections by means of a small illustrative example. The model used to perform our investigations is the well-known Klein's Model I, which gives a very simplified description of the U.S. economy (for the period 1921–1941) consisting of three behavioral equations and three identities.

In its original form the behavioral relations of the model are estimated as follows:

Consumption:

$$\begin{aligned} C = \underset{(4.224)}{16.2365} + \underset{(2.113)}{0.1929}P + \underset{(0.992)}{0.0899}P_{-1} \\ + \underset{(19.925)}{0.7962}(WP + WG)_{-1} + u_G, \end{aligned} \tag{27}$$

Investment:

$$I = \underset{(6.933)}{10.1152} + \underset{(4.941)}{0.4797}P + \underset{(3.302)}{0.3330}P_{-1}$$

$$- \underset{(-4.183)}{0.1117}K_{-1} + u_I, \tag{28}$$

Private wage:

$$WP = \underset{(0.032)}{0.0638} + \underset{(13.571)}{0.4394}(Y + T - WG)$$

$$+ \underset{(3.906)}{0.1461}(Y + T - WG)_{-1} + \underset{(4.085)}{0.1302}(t - 1920)$$

$$+ u_{WP}. \tag{29}$$

The remaining three equations are identities:

Product:

$$Y + T = C + I + G, \tag{30}$$

Income:

$$Y = WP + WG + P, \tag{31}$$

Capital:

$$K = I + K_{-1}. \tag{32}$$

The method of ordinary least squares was applied to obtain numerical estimates for the parameters. The numbers in parentheses below the parameters denote the t-values.

The equations of the model may call for brief comments. The first equation relates linearly total consumption (C) to current lagged profits (P and P_{-1}), with wage ($WP + WG$) being the private and government wage bills, respectively. The second equation represents net investment (I) as a linear function of current and lagged profits, and capital stock at the beginning of the year (K_{-1}). The third equation is the demand for labor function and describes the private wage bill in terms of national income (Y) plus business taxes (T) but excluding the wages paid by the government, the same variable lagged one year, $(Y + T - WG)_{-1}$, and time t measured in calendar years. The remaining three equations are identities, where G stands for the government expenditure and K for the capital stock. The model has six endogenous variables, C, P, WP, I, Y, and K, which are explained by using seven lagged endogenous and exogenous variables, P_{-1}, K_{-1}, $(Y + T - WG)_{-1}$, t, T, G, and WG. All the variables of the model except the time variable t are measured in billions of 1934 U.S. dollars per year. u_C, u_I, and u_{WP} are the disturbance terms.

5.2. Coordinated Policy

It is assumed that the decentralization of the model under consideration is implemented by the two control agencies consisting of public (government) and private sectors. The control (instrument) variables of a government agency are *WG*, *G*, and *T*; that of a private agency is *I* only. Further, the target variables for each agency are *C*, *WP*, *Y*, and *P*. The verifications of the proposed methods are applied using figures for the U.S. economy over the years 1921–1940.

The constraint equations are assumed to be given as follows:

$$C(t) = a_0 + a_1P(t) + a_2P(t-1) + a_3(WP(t) + WG(t)), \qquad (33)$$

$$WP(t) = c_0 + c_1E(t) + c_2E(t-1) + c_3\mathrm{TIME}(t), \qquad (34)$$

$$Y(t) = C(t) + I(t) + G(t) - T(t), \qquad (35)$$

$$P(t) = Y(t) - WP(t) + WG(t), \qquad (36)$$

$$K(t) = I(t) + K(t-1), \qquad (37)$$

$$E(t) = Y(t) + T(t) - WG(t), \qquad (38)$$

$$\mathrm{TIME} = t - 1920, \qquad (39)$$

where the a's, and b's, and c's denote the parameters in (27), (28), and (29), respectively. The control vector is $x' = (WG, G, T, I)$, and the target vector is $y' = (C, WP, Y, P)$. The target vector under the coordinated policy is assumed to be $y^{*\prime} = (C.5, WP.5, Y.5, P.5)$, where, e.g., $C.5$ means that C is kept at 5 percent more than the observation value of C. Therefore, we can write

$$C.5(t) = 1.05 \times C.0(t).$$

The same definitions are applied to the other variables.

We consider the problem to minimize the following quadratic loss

$$\phi = \sum_{t=1291}^{1940} [y(t) - y^*(t)]'Q(t)[y(t) - y^*(t)] \qquad (40)$$

subject to the dynamic constraints (33)–(39), where

$$Q(t) = \mathrm{diag}\,[C.5(t)^{-2}, WP.(t)^{-2}, Y.5(t)^{-2}, P.5(t)^{-2}].$$

The optimal control problem has been computed employing the inverse final test (*X*-test) developed in the POEM (*p*olicy *o*ptimization using *e*conometric *m*odels) algorithm (for the discussions in detail, see Refs. 13, 14).

For the final test (*Y*–test) of the model, the errors for each target variable based on the desired target value are given as *C*, 5.644%; *WP*,

5.692%; Y, 5.526%; P, 7.382%; the error of the whole target variable is 6.109%. On the other hand, for the inverse final test, these errors are given as C, 1.213%; WP, 0.015%; Y, 1.046% P, 0.210%; the error of the whole target variable is 0.808%. It is easily verified from the preceding results that the coordination policy is 99.192% controllable when the target level is raised 5%.

5.3. Uncoordinated Policy: The Government Sector

We now assume that the government agency is strong in policy implementation but the private agency is weak. Under such situations, we consider first the optimization behavior for the government agency. Let the instrument, target variables, and the desired target values for the government agency be denoted by

$$x.G' = (WG, G, T), \qquad y.G' = (WP, Y), \qquad y.G^{*\prime} = (WP.7, Y.7),$$

respectively, i.e., the government agency assumes that the target level is raised 5%. Suppose that the quadratic loss function is

$$\phi.G = \sum_{t=1921}^{1940} [y.G(t) - y.G^*(t)]' Q.G(t)[y.G(t) - y.G^*(t)], \quad (41)$$

where $Q.G(t) = \text{diag}[WP.7(t)^{-2}, Y.7(t)^{-2}]$. Further, the prediction of the government agency is assumed to be $x.P(t) = P.0(t)$.

Under these assumptions, the problem is to find the optimal control solutions to minimize (41) subject to (33)–(39). For the final test, the errors of each variable are as follows: WP, 7.245%; Y, 7.147%; the error of the whole variable is 7.194%. Similarly, for the inverse test, the errors of WP, Y, and the whole variable are 0.002789%, 0.001521%, and 0.00246%, respectively. It is then shown that the degree of controllability for the government agency is 99.99754%, when the target level is kept at 7%.

5.4. Uncoordinated Policy: The Private Sector

Second, we consider the optimization behavior for the private agency. Let each variable of the private agency be defined by

$$x.P = I, \qquad y.P' = (C, WP, P), \qquad y.P^{*\prime} = (C.3, WP.3, P.3),$$

respectively. Suppose that the quadratic loss function is measured by

$$\phi.P = \sum_{t=1921}^{1940} [y.P(t) - y.P^*(t)]' Q.P(t)[y.P(t) - y.P^*(t)] \quad (42)$$

where

$$Q.P(t) = \text{diag}\,[C.3(t)^{-2}, WP.3(t)^{-2}, P.3(t)^{-2}]$$
$$x.G = (WG.0, G.0, T.0).$$

As was seen in Section 5.4, the errors of each variables based on the desired value can also be computed for both tests. For the final test, the errors of *C*, *WP*, *P*, and the whole variable are 4.142%, 4.271%, 6.350%, and 5.024%, respectively. Similarly, for the inverse test, the errors of *C*, *WP*, *P*, and the whole variable are 1.939%, 2.896%, 1.763%, and 2.255%, respectively. It is observed from the preceding results that the degree of controllability for the private agency is 97.745%, when the target level is kept at 3%.

5.5. Comments on the Results

Finally, we consider some results for the case when the uncoordinated policies of both sectors are performed as shown in Sections 5.3 and 5.4. Now, let the desired target level for the system be kept at 5%, the same as we assumed in the coordinated policy case of Section 5.2. In this case, for the final test, the errors of *C*, *WP*, *Y*, *P*, and the whole variable are 0.665%, 3.719%, 4.440%, 9.833%, and 5.716%, respectively.

5.6. The Bias of the Decentralized Policy Rule

Under the coordinated policy, each bias observed from the system is computed as follows: *C*, 1.213%; *WP*, 0.015%; *Y*, 1.046%; *P*, 0.210%; the whole variable, 0.808%. On the other hand, each bias under the uncoordinated policy is also given as *C*, 0.665%; *WP*, 3.719%; *Y*, 4.440%; *P*, 9.833%; and the whole variable, 5.716%. All these numerical values denote the figures of the average percentage over the years 1921–1940. Each bias at every year observed from the system is graphically illustrated in Figs. 1–4.

For example, *C.CO.S* and *C.NC.S* in Fig. 1 denote, respectively, the errors of *C* (consumption) under the coordinated and the uncoordinated policies, when the target level is kept at 5%. The computation of *C.CO.S*(t) is given as

$$C.CO.S(t) = 100 \times \frac{C.CO(t) - C.5(t)}{|C.5(t)|}. \tag{43}$$

Under the uncoordinated policy, each bias observed from the government agency can also be computed. Figure 5 describes the bias of private wage observed from the government agency. For example, the

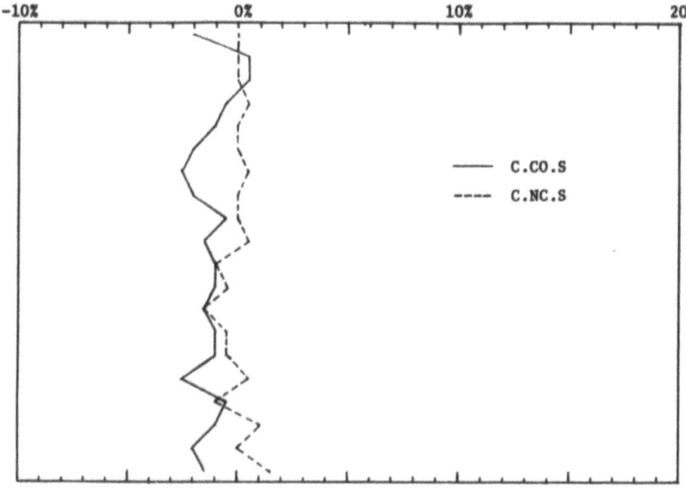

Figure 1. The bias of consumption observed from the system.

error of private wage in Fig. 5 is also computed based on the target level for the government agency. $WP.NC.G(t)$ is computed as

$$WP.NC.G(t) = 100 \times \frac{WP.NC(t) - WP.7(t)}{|WP.7(t)|}. \tag{44}$$

Under the uncoordinated policy, the bias of private wage observed from the private agency is illustrated in Fig. 6. The error denotes the one

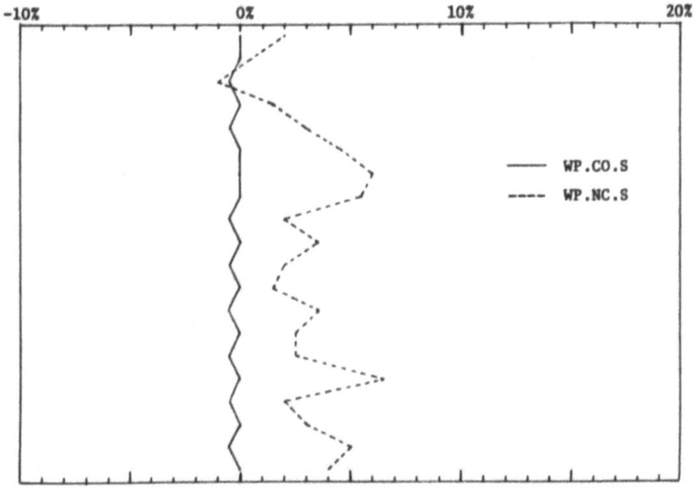

Figure 2. The bias of private wage observed from the system.

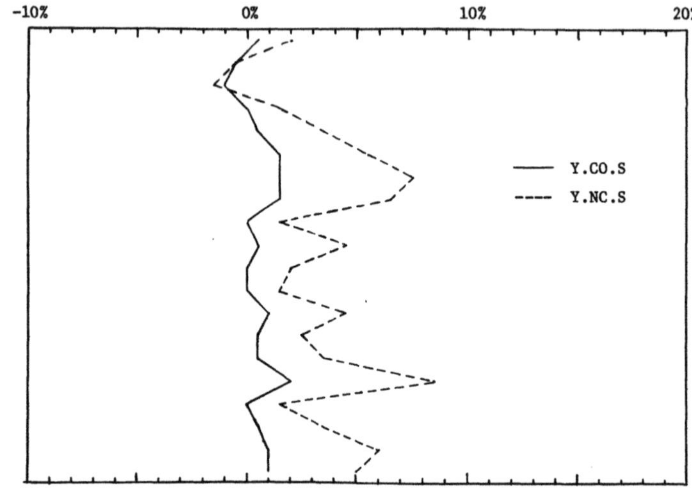

Figure 3. The bias of national income observed from the system.

based on the target level for the private agency. For example, the computation of $WP.NC.P(t)$ in Fig. 6 is given by

$$WP.NC.P(t) = 100 \times \frac{WP.NC(t) - WP.3(t)}{|WP.3(t)|}. \tag{45}$$

As seen in Figs. 5 and 6, each bias observed from the government agency is less than that observed from the private agency. The reasons are

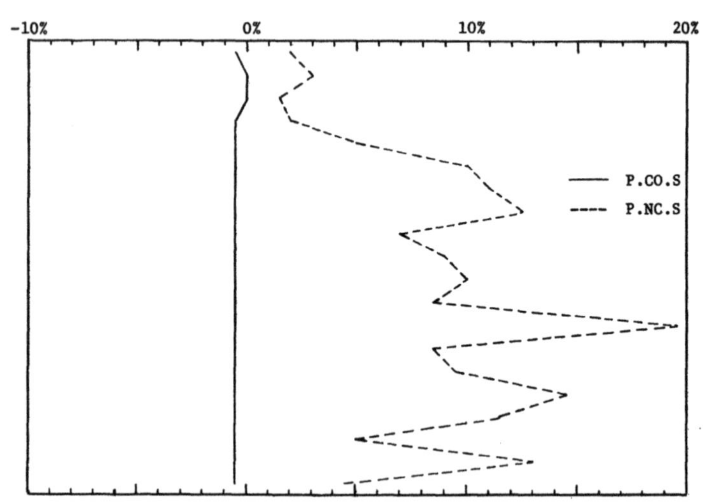

Figure 4. The bias of profits observed from the system.

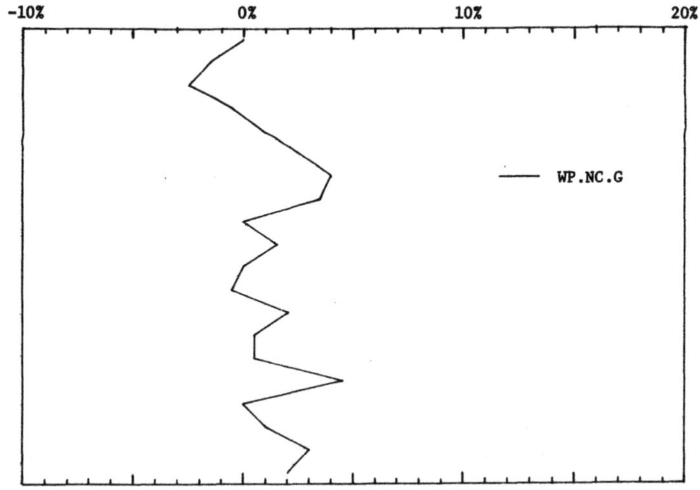

Figure 5. The bias of private wage observed from the government agency.

that more policy instruments are generally available to the government agency than to the private agency and that the degree of controllability for the government is higher. It is observed from Figs 1–4 that the level of the whole target variable under the uncoordinated policy is higher than that of the target variable for the system. This is not without reason. First, the level (7%, say) of the target variable for the government agency is higher than that (5%, say) of the target variable for the system. Second,

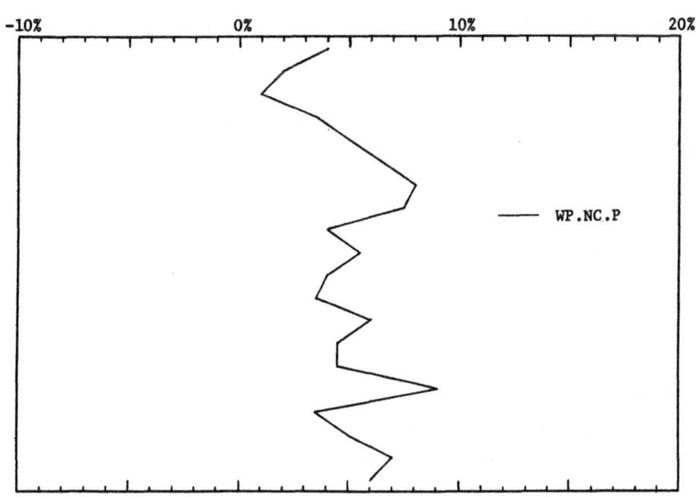

Figure 6. The bias of private wage observed from the private agency.

as observed before, the controllability for the government agency is higher compared with that for the private agency. Third, the policy instruments by one agency are obtained from compensation for these by another agency, since both the government agency and the private agency underestimate the policy instrument (0%, say) by the other agency. Therefore, each bias under the uncoordinated policy becomes small, if each policy instrument by the other agency is estimated at a reasonable level.

6. Concluding Remarks

The decentralization constraints make control of the large-scale systems realistic. The study of such large-scale systems is dominated by a concern for optimality. As was mentioned in Section 1, the computational effort required for calculating optimal decision rules for large-scale systems with emphasis on decentralized control strategies is considerably greater than that for centralized decision making. In this paper the concern is stability and stabilizability, rather than optimality. Decentralized stabilization policies were investigated using a very simple macroeconometric model. It should be noted that the results obtained here are restricted by the model presented in Section 5.

Some possible directions for future research will be indicated. The analytical framework proposed here is well suited for dealing with the more general problem of decentralized stabilization policies in the international economy. In this case, for example, the model is treated as a dynamic nonzero-sum game of two countries faced by the independence of the fortunes. Further, this framework is applied to analyze decentralization versus coordination of monetary and fiscal policies in the interdependent economy and will also stimulate others to investigate different types of games employing actual LINK econometric models (Ref. 15).

ACKNOWLEDGMENT

The author thanks Yukio Uchida for his help with the computations.

References

1. TINBERGEN, J., *On the Theory of Economic Policy*, North-Holland, Amsterdam, 1952.
2. THEIL, H., *Optimal Decision Rules for Government and Industry*, North-Holland, Amsterdam, 1964.
3. WITSENHAUSEN, H. S., A counterexample in stochastic optimal control, *SIAM Journal of Control*, Vol. 6, pp. 131–147, 1968.

4. "Special Issue on Large-Scale Systems and Decentralized Control," *IEEE Transactions on Automatic Control,* Vol. AC-23, pp. 105–371, 1978.
5. Ho, Y. C. and Chu, K. C., Team decision theory and information structures in optimal control problems. Part I, *IEEE Transactions on Automatic Control,* Vol. AC-17, pp. 15–22, 1972.
6. Sandell, N. and Athans, M., Solution of some nonclassical LQG stochastic decision problems, *IEEE Transactions on Automatic Control,* Vol. AC-19, pp. 108–116, 1974.
7. McFadden, D., On the controllability of decentralized macroeconomic systems: The assignment problem," in *Mathematical Systems Theory and Applications.* Edited by H. W. Kuhn and P. Szegö, Springer-Verlag, Berlin/Heidelberg/New York, 1967.
8. Aoki, M., On decentralized stabilization policies and dynamic assignment problems, *Journal of International Economics,* Vol. 6, pp. 143–171, 1976.
9. Pindyck, R. S., Optimal economic stabilization policies under decentralized control and conflicting objectives, *IEEE Transactions on Automatic Control,* Vol. AC-22, pp. 517–530, 1977.
10. Klein, L. R., *Economic Fluctuations in the United States, 1921–1941,* John Wiley and Sons, New York, 1950.
11. Fisher, M. E. and Fuller, A. T., On the stabilization of matrices and the convergence of linear iterative processes, *Proceedings of the Cambridge Philosophical Society,* Vol. 54, pp. 417–425, 1958.
12. Chow, G. C., *Analysis and Control of Dynamic Economic Systems,* John Wiley and Sons, New York, 1975.
13. Myoken, H. and Uchida, Y., System modelling for interconnected dynamic economy and decentralized optimal control, in *Optimization Techniques, Part II, Lecture Notes in Control and Information Sciences.* Edited by J. Stoer, Springer-Verlag, Berlin/Heidelberg/New York, 1978.
14. Myoken, H., Sadamichi, H., and Uchida, Y., Decentralized stabilization and regulation in large-scale macroeconomic–environmental models, and conflicting objectives, *Proceedings of IFAC Workshop on Urban, Regional and National Planning,* Pergamon Press, Oxford/New York, 1977.
15. Ball, R. J. (ed), *The International Linkage of National Economic Models,* North-Holland, Amsterdam, 1973.

15

A Minimum Principle for Decentralized Stochastic Control Problems

PRAVIN VARAIYA AND JEAN WALRAND

1. Introduction

The notion of the "state" of a dynamical system has turned out, in retrospect, to be the single most important concept in system theory, and the representation of systems in state space form is now almost universally accepted as the correct framework for posing questions of control and for formulating and analyzing proposed solutions. The power of the state concept derives from two properties. First, the current value of the state is a sufficient (and "minimal") statistic of all the past history of the system that is relevant for *predicting* future behavior. Second, the system's optimal future behavior depends only on the current state, so it is a sufficient statistic for *controlling* future behavior in an optimal manner. As is well known, however, these properties can be exploited only when the state is directly observed—the so-called complete observation case—and so the utility of the state concept is reduced considerably when it can be observed only indirectly—the partial observation case.

In the case of partial observation one may instead use the probability distribution of the current state conditioned on the available observations. This conditional distribution is evidently computable from the available

PRAVIN VARAIYA and JEAN WALRAND • Department of Electrical Engineering and Computer Sciences, and the Electronics Research Laboratory, University of California, Berkeley, California. Present address of J. W.: School of Electrical Engineering Cornell University, Ithaca, New York.

information, and it enjoys these two properties; therefore from a theoretical view it can be satisfactorily regarded as a state.† Unfortunately, the transformation of the observations into this conditional distribution generally imposes a heavy computational burden and does not lead to any data reduction. Hence, in applications, the complexity of the problem of optimal control with partial observation manifests itself as a computational difficulty.

Another useful way to appreciate this complexity is to note that when the state is not directly observed, it must be estimated from the available observations. The uncertainty associated with the state estimate will generally depend on the control inputs. Thus, in addition to guiding system behavior, the control inputs also acquire an additional, "dual" function of "learning."‡ Since the optimal law must implicitly balance these two often conflicting functions, its structure is more complex than when the state is observed directly so that the learning function is absent.

Several procedures have been proposed to design control laws for the partial observation case.§ While these procedures demand considerably less computational effort than is needed for calculating the optimal law, it is not known how well the resulting nonoptimal laws perform in comparison with the truly optimal law.¶ In the next section yet another class of nonoptimal control laws is proposed. A particular control law from this class is *consistent* in the sense that if at any time t it is assumed that for all other times $s \neq t$ the law is not to be changed, then the performance cannot be improved by deviating from the law at time t. Consistent laws are therefore "locally-in-time" optimal, and this optimality is characterized by a minimum principle.‖

When the observations and choice of control inputs are shared

† This statement is proved by Striebel (Ref. 1). Another alternative proposed by Witsenhausen (Ref. 2), is to define as a new state the sequence of all past "old" states and inputs. Then the control law is memoryless.

‡ The first systematic treatment of this dual function was carried out by Feldbaum (Ref. 3). Further investigation was made by Bar-Shalom and Tse, see Ref. 4 and the references cited there. A thorough investigation of adaptive control for macroeconomic stabilization is given by Kendrick (Ref. 5).

§ In (Ref. 4) an attempt is made to locate these procedures in a taxonomic scheme based on the extent to which an account of the learning function is undertaken.

¶ When the control is to be exercised over an infinite time interval and the performance of control is taken to be the average cost per unit time, control laws—so-called self-tuning regulators and their extensions—have been proposed that are both computationally attractive and known to be optimal under certain conditions. For the case of linear quadratic Gaussian problems see the survey by Wittenmark (Ref. 6); in a very remarkable paper (Ref. 7), Mandl shows the optimality of these self-tuning controllers for a class of finite-state Markov problems.

‖ This notion of consistency and the minimum principle based on it are due to Elliott (Ref. 8) who studied controlled Ito processes.

among several controllers—the "team" decision problem—it seems that the notion of state has little theoretical or practical significance. Indeed it appears that the state concept presupposes the perspective of a single controller: the state summarizes the information available to a single observer–controller. Since within a team there is a multiplicity of perspectives, there is a different information state for each controller. Moreover, each controller's observation is affected by the control inputs chosen by all the controllers. Hence the optimal control law must now balance the task of guiding the system against what it can learn about the state from its own observations as well as against what the other controllers can learn about the state from their observations. These may be termed triple-control problems.† The "consistent" laws for the single controller problem can be extended to team problems, the local-in-time optimality being weakened further to a local-in-time-and-controller optimality: a consistent law is now one that cannot be improved on by any single controller deviating from it at any single time instant. In Section 3 this characterizing condition is also presented as a minimum principle.

2. Minimum Principle for a Single Controller

The behavior of a system is governed by the equation

$$x_t = f_t(x_{t-1}, u_t, v_t), \qquad t = 1, \ldots, T, \tag{1}$$

where at time t, x_t is the "state," u_t is the control input, and v_t is the random disturbance. The primitive random variables

$$\omega = (x_0, v_1, \ldots, v_T)$$

are all independent. At each time t the controller observes only a subvector‡ y_{t-1} of the state x_{t-1}, and the controller must select the input u_t from a prespecified set U_t. This selection is based on the information available up to time t, namely

$$y^{t-1} = \{y_s \mid 0 \leq s \leq t - 1\}. \tag{2}$$

An (admissible) control *law* then is any set of functions $\gamma = \{\gamma_t \mid 1 \leq t \leq T\}$ such that

$$\gamma_t \colon y^{t-1} \mapsto u_t = \gamma_t(y^{t-1}) \in U_t.$$

† The effect of one controller's action on the observations made by another is called signaling in the theory of games (Ref. 9). In control theory signaling was explicitly introduced by Sandell and Athans (Ref. 10) and has been fairly extensively investigated by Ho, Kastner, and Wong (Ref. 11).

‡ It is more customary to split x_t into a state vector z_t and an observation vector y_t, but (1) is more compact.

Γ is the set of all laws. (Here and later many difficult technical considerations of measurability are ignored.)

In analogy with the notation (2) it will prove convenient to adopt

$$x^t = \{x_s \mid 0 \le s \le t\}, \qquad u^t = \{u_s \mid 1 \le s \le t\},$$
$$\gamma^t = \{\gamma_s \mid 1 \le s \le t\}. \quad (3)$$

Now fix a law γ. Recursive substitution of $u_t = \gamma_t(y^{t-1})$ into (1) will generate the state and control processes $\{x_t\}$, $\{u_t\}$. These are sequences of random variables whose probability distribution is determined by γ. (However, to avoid cumbersome notation we do not indicate explicitly the dependence of x_t on γ.) E_γ denotes the expectation operator with respect to this probability distribution. Thus, for example, $E_\gamma\{h(x_t)\}$ is the expectation of $h(x_t)$ where x_t is the random variable denoting the state at time t when the system is controlled by the law γ. It is obvious but important to note that the distribution of x^t, y^t, u^t is determined solely by $\gamma_1, \ldots, \gamma_t$ (and the distribution of ω) and does not depend on $\gamma_{t+1}, \ldots, \gamma_T$.

The performance of any law γ is evaluated by the cost

$$J(\gamma) = E_\gamma \sum_1^T c_t^\gamma = E_\gamma \sum_1^T c_t(x_t, u_t), \quad (4)$$

where the c_t are prespecified functions. The *value process corresponding to* γ is the sequence of random variables

$$V_t^\gamma = V_t^\gamma(y^t) = \text{Inf}\left[E_\eta\left\{\sum_{t+1}^T c_s^\eta \mid y^t\right\} \,\middle|\, \eta \in \Gamma, \eta_1 = \gamma_1, \ldots, \eta_t = \gamma_t\right]. \quad (5)$$

In words, V_t^γ is the minimum future cost conditional on y^t and assuming that the law γ is followed up to time t. Evidently,

$$V_0 = \text{Inf}\left[E\left\{\sum_1^T c_s^\gamma \mid y^0\right\} \,\middle|\, \gamma \in \Gamma\right]$$

is independent of γ and

$$J^* = EV_0$$

is the minimum possible cost. The law γ is said to be optimal if $J(\gamma) = J^*$. The proof of the next result requires a straightforward application of dynamic programming techniques. It is given in the appendix.

Theorem 2.1 (Principle of Optimality). For every γ

$$V_t^\gamma \le E_\gamma\{c_{t+1}^\gamma + V_{t+1}^\gamma \mid y^t\}, \qquad t = 0, 1, \ldots, T. \quad (6)$$

Moreover, γ is optimal if and only if equality holds for every t.

This result has been proved for Brownian motion disturbance by Rishel (Ref. 12), Davis and Varaiya (Ref. 13), and in a very general setting by Striebel (Ref. 14). An economic interpretation of (6) can be given by defining

$$\tilde{V}_t^\gamma = E_\gamma\left\{\sum_1^t c_s \mid y^t\right\} + V_t^\gamma,$$

the minimum *total* cost given the observations up to t and using γ up to t. Then (6) can be rewritten as

$$\tilde{V}_t^\gamma \leq E_\gamma\{\tilde{V}_{t+1}^\gamma \mid y^t\}$$

with equality holding for all t if and only if t is optimal. That is, $\tilde{V}^\gamma$ is a supermartingale if γ is not optimal and a martingale if it is optimal. This extends an observation of Samuelson (Ref. 15) to the effect that the price of an asset, if it is correctly evaluated, must form a martingale.

The practical usefulness of this result is slight. For purposes of comparison consider first the case of complete observation, i.e., $y_t = x_t$. In this case it can be shown that $V_t^\gamma(y^t) = V_t^\gamma(x^t)$ does not depend on γ. Moreover, there exist functions $V_t(x_t)$ such that $V_t^\gamma(x^t) = V_t(x_t)$, and these can be computed by backwards recursion:

$$V_T(x_T) \equiv 0,$$

$$V_t(x_t) \equiv \operatorname{Inf}\left[E\{c_{t+1}(x_{t+1}, u_{t+1}) + V_{t+1}(x_{t+1}) \mid x_t\} \mid u_{t+1} \in U_{t+1}\right]. \tag{7}$$

This computation also yields the optimal law $u_{t+1} = \gamma_{t+1}^*(x_t)$ as the one where the infimum in (7) is achieved.

In the partial observation case V_t^γ depends on γ. Worse yet, V_t^γ cannot even be computed recursively in a form similar to (7). To see this, introduce the notation γ/η_t to signify, for any two laws γ and η, the law

$$\gamma/\eta_t = \{\gamma_1, \ldots, \gamma_{t-1}, \eta_t, \gamma_{t+1}, \ldots, \gamma_T\}. \tag{8}$$

Now, for any law γ, consider the sequence of functions $W_t^\gamma(y^t)$ defined by analogy with (7):

$$W_T^\gamma(y^T) \equiv 0,$$
$$W_t^\gamma(y^t) \equiv \operatorname{Inf}\left[E_{\gamma/\eta_{t+1}}\{c_{t+1}^\eta + W_{t+1}^\gamma(y^{t+1}) \mid y^t\} \mid \eta \in \Lambda\right]. \tag{9}$$

It may appear at first sight that W_t^γ and the value process V_t^γ should be equal, but it is not so. Second, suppose that the infimum in (9) is achieved at η. It may be conjectured that η is then the optimal law. Unfortunately, this conjecture is also false. Suppose further that the infimum in (9) is achieved at γ itself; even then it does *not* follow that γ is optimal.

Theorem 2.1 cannot even be used to decide whether or not a particular law γ is optimal, for γ is optimal if and only if

$$V_t^\gamma = E_\gamma\left\{\sum_{t+1}^{T} c_s \mid y^t\right\}.$$

The right-hand side can evidently be computed if we know γ; but, as we have seen, there is no simple procedure for calculating the left-hand side. It turns out, however, that with a small change we can obtain the following verification theorem.

Definition 2.1. γ is said to be *consistent* or *locally-in-time optimal* if for all η and t

$$J(\gamma) \leq J(\gamma/\eta_t).$$

Theorem 2.2. Let γ be any law, and define the functions $W(t, x^t)$, $t = 0, \ldots, T$, by $W(T, x^T) \equiv 0$,

$$W(t, x^t) = E_\gamma\left\{\sum_{t+1}^{T} c_s^\gamma \mid x^t\right\} = E_\gamma\{c_{t+1} + W(t+1, x^{t+1}) \mid x^t\}. \quad (10)$$

Then γ is consistent if and only if for all η and t

$$E_{\gamma|\eta_{t+1}}\{c_{t+1}^\eta + W(t+1, x^{t+1}) - W(t, x^t) \mid y^t\} \geq 0. \quad (11)$$

The proof is given in the appendix.

Keeping with the notation introduced before, define the "Hamiltonian" functions

$$H_{t+1}(x^t, u_{t+1}, v_{t+1}) = W(t+1, x^t, f_{t+1}(x_t, u_{t+1}, v_{t+1})) - W(t, x^t),$$

$$t = 0, \ldots, T-1.$$

In terms of these functions Theorem 2.2 can be restated as a minimum principle [$W_0(x_0) = W(0, x_0)$, below.]

Corollary 2.1 (Minimum Principle). Suppose γ is optimal. Then there exist functions $H_t(x^{t-1}, u_t, v_t)$ and $W_0(x_0)$ so that

(i) $$\sum_{1}^{T} H_t(x^{t-1}, \eta_t(y^{t-1}), v_t) + W_0(x_0) \equiv 0 \qquad \text{for all } \eta \text{ in } \Gamma;$$

(ii) $$\min_{\eta\in\Gamma} E_{\gamma/\eta_t}\{c_t^\eta + H_t(x^{t-1}, \eta_t(y^{t-1}), v_t) \mid y^{t-1}\} = 0,$$

and the minimum is achieved at γ.

Theorem 2.2 or Corollary 2.1 is of course only a necessary condition for (global) optimality. But it can at least be used constructively to partially verify whether a specific law γ is optimal. A particular law that satisfies the condition may be improved by altering it at two or more time instants. This interpretation suggests a sufficient condition for optimality. For γ, η in Γ, let $\gamma \mid \eta^t$ denote the law

$$\gamma/\eta^t = \{\eta_1, \ldots, \eta_t, \gamma_{t+1}, \ldots \gamma_T\}. \tag{12}$$

Theorem 2.3. γ is optimal if there exist functions $W(t, x^t)$, $t = 0, \ldots, T$, with $W(T, x^T) \equiv 0$ such that for all η in Γ

$$\begin{aligned} E_\eta\{c_{t+1}^\eta + W(t+1, x^{t+1}) - W(t, x^t) \mid y^t\} \\ \geq E_{\gamma/\eta^t}\{c_{t+1}^\gamma + W(t+1, x^{t+1}) - W(t, x^t) \mid y^t\} = 0. \end{aligned} \tag{13}$$

Proof. The expression (13) implies, in particular, that

$$E_\eta\{W(t, x^t) \mid y^t\} \leq E_\eta\{c_{t+1}^\eta + W(t+1, x^{t+1}) \mid y^t\},$$

and so

$$E_\eta\{W(t, x^t) \mid y^t\} \leq E_\eta\left\{\sum_{t+1}^{T} c_s^\eta \mid y^t\right\}.$$

Therefore

$$EW(0, x_0) = E_\eta W(0, x_0) \leq E_\eta\left\{\sum_1^T c_t^\eta\right\} = J(\eta).$$

On the other hand, use of the equality in (13) shows that

$$EW(0, x_0) = J(\gamma),$$

proving that γ is optimal. □

Condition (13) is highly restrictive. It asserts that the functions $W(t, x_t)$, which correspond to the optimal law γ, are such that γ appears to be a superior law even if these are evaluated under the assumption that η was used in the past. Intuitively speaking, the condition implies that the control function (at least when the optimum law is used) "dominates" the learning function. When can this happen? One instance is the case of complete information, as already noted. Another instance is the so-called feedforward case when the state dynamics does not depend on u_t, i.e., instead of (1) one has $x_t = f_{t-1}(x_{t-1}, v_t)$. This is a common formulation in the portfolio selection problem.† A third instance is the case when the optimal control makes future state values independent of past observations.

† The literature on portfolio selection problems is large. Two seminal articles are Refs. 16, 17. Benes (Ref. 18) has also considered a problem in which the dynamics are unaffected by the control.

3. A Minimum Principle for Decentralized Control

Now consider a system whose behavior is governed by

$$x_t = f_t(x_t, u_t, v_t) = f_y(x_t, u_t^1, \ldots, u_t^K, v_t), \qquad t = 1, \ldots, T, \tag{14}$$

where at each time t, u_t^k is the control input selected by the kth controller. This selection must be made from a prespecified set U_t^k and can depend on the information

$$y^{k,t-1} = \{y_s^k \mid 0 \le s \le t-1\}.$$

Here y_t^k is a subvector of the state x_t. Thus the system is controlled by a "team" of K controllers, each of which has its own observation. A team control law is now any set of functions $\gamma = \{\gamma_t^k \mid 1 \le t \le T, 1 \le k \le K\}$ such that

$$\gamma_t^k : y^{k,t-1} \mapsto u_t^k = \gamma_t^k(y^{k,t-1}) \in U_t^k.$$

The team's objective is to find a law γ that minimizes

$$J(\gamma) = E_\gamma \sum_1^T c_t^\gamma = E_\gamma \sum_1^T c_t(x_t, u_t^1, \ldots, u_t^K).$$

In analogy with (8), for any two laws γ and η, and any k, γ/η_t^k denotes the law $\alpha = \{\alpha_s^j\}$ where

$$\alpha_s^j = \gamma_s^j \quad \text{if} \quad (j, s) \ne (k, t), \alpha_t^k = \eta_t^k.$$

Definition 3.1. γ is said to be *consistent* or *locally-in-time and by-person optimal* if for all η, k, and t

$$J(\gamma) \le J(\gamma \mid \eta_t^k).$$

Theorem 3.1. Let γ be any law and define the functions $W(t, x^t)$, $t = 0, \ldots, T$, by $W(T, x^T) \equiv 0$,

$$W(t, x^t) = E_\gamma\left\{\sum_{t+1}^T c_s^\gamma \mid x^t\right\} = E_\gamma\{c_{t+1}^\gamma + W(t+1, x^{t+1}) \mid x^t\}.$$

Then γ is consistent if and only if for all η, k, and t

$$E_{\gamma/\eta_{t+1}^k}\{c_{t+1}^{\eta^k} + W(t+1, x^{t+1}) - W(t, x^t) \mid y^{k,t}\} \ge 0.$$

{Here $c_{t+1}^{\eta^k} = c_{t+1}[x_{t+1}, \gamma_{t+1}^1(y^{1,t}), \ldots, \eta_{t+1}^k(y^{k,t}), \ldots, y_{t+1}^K(y^{K,t})]$.}

Proof. Follows immediately from the preceding definition and Theorem 2.2. □

The condition has some intuitive appeal. The difference $W(t, x^t) - W(t + 1, x^{t+1})$, which is the reduction in the future cost over one period, serves as a partial decentralizing signal. Although it cannot be computed by any controller (since x_t is not observed), consistency requires that for each controller k the expected value of this signal given the observation $y^{k,t}$ should not exceed the expected value of the one-period cost, c_{t+1}, no matter which law η_{t+1}^k is used.

In analogy with (12), for any γ, η, and k, let $\gamma/\eta^{k,t}$ denote the law $\alpha = \{\alpha_s^j\}$, where

$$\alpha_s^j = \gamma_s^j \quad \text{if } j \neq k; \qquad \alpha_s^k = \eta_s^k \quad \text{if } s \leq t; \qquad \alpha_s^k = \gamma_s^k \quad \text{if } s \geq t + 1.$$

Also let γ/η^k denote the law $\alpha = \{\alpha_s^j\}$, where

$$\alpha_s^j = \gamma_s^j \quad \text{if } j \neq k, \qquad \alpha_s^k = \eta_s^k.$$

γ is said to be *person-by-person optimal* if $J(\gamma) \leq J(\gamma/\eta^k)$ for all η, k. The next result follows from Theorem 2.3.

Theorem 3.2. γ is person-by-person optimal if there exist functions $W(t, x^t)$, $t = 0, \ldots, T$, with $W(T, x^T) \equiv 0$ such that for all η, k

$$\begin{aligned} E_{\gamma/\eta^k}\{c_{t+1}^{\eta^k} + W(t + 1, x^{t+1}) - W(t, x^t) \mid y^{k,\eta}\} \\ \geq E_{\gamma/\eta^{k,t}}\{c_{t+1}^{\gamma} + W(t + 1, x^{t+1}) - W(t, x^t) \mid y^{k,\eta}\} = 0. \end{aligned}$$

4. Concluding Remarks

Control of systems that are only partially observed is difficult because the control action influences learning in addition to guiding the state trajectory. The difficulty is compounded when control is decentralized. It is possible to extend the optimality conditions derived for the complete observation case to the case of partial observations by separating the two functions and evaluating the cost incurred by one feedback law on the basis of information collected under another law. Using this device the paper reports one necessary and one sufficient condition for optimality. The necessary condition characterizes team control laws that are locally-in-time and by person optimal.

ACKNOWLEDGMENTS

This research was sponsored by the Joint Services Electronics Program Contract F44620-76-C-0100. The authors are indebted to Professor Robert Elliott for very helpful discussions.

Appendix

Proof of Theorem 2.1. For any γ let

$$J_t^\gamma = E_\gamma\left\{\sum_{t+1}^{T} c_s^\gamma \mid y^t\right\},$$

so that the value process corresponding to γ is

$$V_t^\gamma = \operatorname{Inf}[J_t^\eta \mid \eta \in \Gamma, \eta^t = \gamma^t]. \tag{A1}$$

(Here Inf means essential infimum.)

Lemma A.1. Fix $\gamma, \varepsilon > 0, t \geq 0$. Then there exists η with $\eta^t = \gamma^t$ such that

$$J_t^\eta \leq V_t^\gamma + \varepsilon, \qquad \text{almost surely.}$$

Proof. Let n_1 be the smallest integer such that there exists η^1 with $\eta^{1t} = \gamma^t$ and

$$P(A_1) \equiv P\{y^t \mid J_t^{\eta^1}(y^t) \leq V_t^\gamma(y^t) + \varepsilon\} \geq n_1^{-1}.$$

(Here and later P denotes the probability measure induced on y^t by γ^t.) Next let $n_2 \geq n_1$ be the smallest integer such that there exists η^2 with $\eta^{2t} = \gamma^t$ and

$$P(A_2) \equiv P(A_2) \equiv P\{y^t \mid y^t \notin A_1, J_t^{\eta^2}(y^t) \leq V_t^\gamma(y^t) + \varepsilon\} \geq n_2^{-1}.$$

Proceed inductively in this way until at step r, laws $\eta^1, \ldots, \eta^r$ with corresponding disjoint sets $A_1, \ldots, A_r$ have been selected. At step $r+1$ let $n_{r+1} \geq n_r$ be the smallest integer such that there exists η^{r+1} with $\eta^{r+1t} = \gamma^t$ and

$$P(A_{r+1}) \equiv P\left\{y^t \mid y^t \notin \bigcup_1^r A_i, J_t^{\eta^{r+1}}(y^t) \leq V_t^\gamma(y^t) + \varepsilon\right\} \geq n_{r+1}^{-1}.$$

If no finite integer n_{r+1} exists, then set $n_{r+1} = n_{r+2} = \cdots = \infty$ and $A_{r+1} = A_{r+2} = \cdots = \emptyset$. Define

$$A = \bigcup_1^\infty A_r.$$

We claim that

$$P(A) = \sum_1^\infty P(A_r) = 1. \tag{A2}$$

Indeed if the assertion were false, then by (A1) there would exist $\tilde{\eta}$ with $\tilde{\eta}^t = \gamma^t$ such that

$$P(\tilde{A}) \equiv P\{y^t \mid y^t \notin A, J_t^{\tilde{\eta}}(y^t) \leq V_t^\gamma(y^t) + \varepsilon\} = \delta > 0.$$

But then there exists r with $n_r^{-1} \geq \delta > n_{r+1}^{-1}$, which contradicts the definition of n_{r+1} in the previous construction. Hence (A2) must hold.

Finally consider $\eta = (\eta_1, \ldots \eta_T)$ with $\eta^t = \gamma^t$, and for $s \geq t + 1$

$$\eta_s(y^{s-1})\begin{cases} = \eta_s^r(y^{s-1}), & \text{if } y^t \in A_r, \\ \in U_s \text{ arbitrary}, & \text{if } y^t \notin A. \end{cases}$$

Evidently η is an admissible law, and

$$J_t^\eta(y^t) = \sum_1^\infty 1_{A_r}(y^t) J_t^{\eta^r}(y^t) \leq V_t^\gamma + \varepsilon, \qquad \text{almost surely.}$$

(Here $1_B(\cdot)$ denotes the indicator function of the set B.) The lemma is proved. □

We turn now to the proof of Theorem 2.1. Fix γ and t. By (A1)

$$\begin{aligned} V_t^\gamma &= \operatorname{Inf}[J_t^\eta \mid \eta \in \Gamma, \eta^t = \gamma^t] \\ &\leqq \operatorname{Inf}[J_t^\eta \mid \eta \in \Gamma, \eta^{t+1} = \gamma^{t+1}] \\ &= \operatorname{Inf}\left[E_\eta\left\{c_{t+1}^\gamma + \sum_{t+2}^T c_s^\eta \mid y^t\right\} \middle| \eta \in \Gamma, \eta^{t+1} = \gamma^{t+1}\right] \\ &= E_\gamma\{c_{t+1}^\gamma \mid y^t\} + \operatorname{Inf}\left[E_\eta\left\{\sum_{t+2}^T c_s^\eta \mid y^t\right\} \middle| \eta \in \Gamma, \eta^{t+1} = \gamma^{t+1}\right]. \end{aligned} \tag{A3}$$

Consider the last term. Let $\varepsilon > 0$. By the preceding lemma, there is η with $\eta^{t+1} = \gamma^{t+1}$ such that

$$J_{t+1}^\eta = E_\eta\left\{\sum_{t+2}^T c_s^\eta \mid y^{t+1}\right\} \leq V_{t+1}^\gamma + \varepsilon, \qquad \text{almost surely,}$$

so that

$$E_\eta\left\{\sum_{t+2}^T c_s^\eta \mid y^t\right\} \leq E_\eta\{V_{t+1}^\gamma \mid y^t\} + \varepsilon,$$

which, together with (A3), implies

$$V_t^\gamma \leq E_\gamma\{c_{t+1}^\gamma \mid y^t\} + E_\eta\{V_{t+1}^\gamma \mid y^t\} + \varepsilon.$$

Since $\eta^{t+1} = \gamma^{t+1}$,

$$E_\eta\{V_{t+1}^\gamma \mid y^t\} = E_\gamma\{V_{t+1}^\gamma \mid y^t\},$$

so that

$$V_t^\gamma \le E_\gamma\{c_{t+1}^\gamma + V_{t+1}^\gamma \mid y^t\}. \tag{A4}$$

Now suppose that γ is such that equality holds in (A4) for all t. Then, summing both sides and taking expectations gives

$$EV_0^\gamma = E_\gamma \sum_1^T c_s^\gamma = J(\gamma),$$

but $EV_0^\gamma = J^*$, so that γ is optimal.

Conversely suppose that (A4) does not hold with equality for t. Then taking expectations and summing both sides gives

$$J^* = EV_0^\gamma < J(\gamma),$$

so that γ cannot be optimal.

Proof of Theorem 2.2. We begin with two preliminary results.

Lemma A.2. γ is consistent if and only if for all η and t

$$E_{\gamma|\eta_{t+1}}\left\{c_{t+1}^\eta + \sum_{t+2}^T c_s^\gamma \mid y^t\right\} - E_\gamma\left\{\sum_{t+1}^T c_s^\gamma \mid y^t\right\} \ge 0, \qquad \text{almost surely.} \tag{A5}$$

Proof. To obtain the sufficiency observe that

$$\begin{aligned} J(\gamma \mid \eta_{t+1}) &= E_\gamma\left\{\sum_1^t c_s^\gamma\right\} + E_{\gamma|\eta_{t+1}}\left\{c_{t+1}^\eta + \sum_{t+2}^T c_s^\gamma\right\} \\ &\le E_\gamma\left\{\sum_1^T c_s^\gamma\right\} = J(\gamma) \qquad \text{[by (A5)]}. \end{aligned}$$

To obtain necessity suppose that there is a y^t-measurable set A with $E_\gamma 1_A > 0$ such that

$$E_\gamma\left\{\sum_{t+1}^T c_s^\gamma \mid y^t\right\} > E_{\gamma|\eta_{t+1}}\left\{c_{t+1}^\eta + \sum_{t+2}^T c_s^\gamma \mid y^t\right\} \qquad \text{for} \quad y^t \in A, \tag{A6}$$

and consider the law $\gamma \mid \xi_{t+1}$, where

$$\xi_{t+1}(y^t) = \begin{cases} \eta_{t+1}(y^t) & \text{if } y^t \in A, \\ \gamma_{t+1} & \text{if } y^t \notin A. \end{cases}$$

Then

$$E_{\gamma|\xi_{t+1}}\left\{c_{t+1}^{\xi} + \sum_{t+2}^{T} c_s^{\gamma} \mid y^t\right\} = 1_A(y^t)E_{\gamma|\eta_{t+1}}\left\{c_{t+1}^{\eta} + \sum_{t+2}^{T} c_s^{\gamma} \mid y^t\right\} + (1 - 1_A(y^t))E_{\gamma}\left\{\sum_{t+1}^{T} c_s^{\gamma} \mid y^t\right\},$$

and so

$$J(\gamma) - J(\gamma \mid \xi_{t+1}) = E_{\gamma}1_A\left[E_{\gamma}\left\{\sum_{t+1}^{T} c_s^{\gamma} \mid y^t\right\} - E_{\gamma|\eta_{t+1}}\left\{c_{t+1}^{\eta} + \sum_{t+2}^{T} c_s^{\gamma} \mid y^t\right\}\right] > 0,$$

by (A6); γ cannot be consistent. □

The following proposition is an immediate consequence of the fact that for any law γ the probability distribution of x^t, y^t, u^t is determined by γ^t (and the distribution of ω) and does not depend on $\gamma_{t+1}, \ldots, \gamma_T$.

Proposition A.1. For any function $h(\omega, x^{t+s})$, and any laws α, β, δ such that $\alpha^{t+s} = (\beta_1, \ldots, \beta_t, \delta_{t+1}, \ldots, \delta_{t+s})$,

$$E_\beta\{E_\delta\{h(\omega, x^{t+s}) \mid x^t\} \mid y^t\} = E_\alpha\{h(\omega, x^{t+s}) \mid y^t\}.$$

We can now complete the proof of Theorem 2.2. Recall that

$$W(t, x^t) = E_{\gamma}\left\{\sum_{t+1}^{T} c_s^{\gamma} \mid x^t\right\},$$

and so, trivially from the proposition (taking $s = 0$),

$$E_{\gamma|\eta_{t+1}}\{W(t, x^t) \mid y^t\} = E_{\gamma}\left\{\sum_{t+1}^{T} c_s^{\gamma} \mid y^t\right\}. \tag{A7}$$

Similarly the proposition yields

$$E_{\gamma|\eta_{t+1}}\{W(t, x^{t+1}) \mid y^{t+1}\} = E_{\gamma|\eta_{t+1}}\left\{E_{\gamma}\left\{\sum_{t+2}^{T} c_s^{\gamma} \mid x^{t+1}\right\} \Bigm| y^{t+t}\right\} = E_{\gamma|\eta_{t+1}}\left\{\sum_{t+2}^{T} c_s^{\gamma} \mid y^{t+1}\right\},$$

so that,

$$E_{\gamma|\eta_{t+1}}\{W(t, x^{t+1}) \mid y^t\} = E_{\gamma|\eta_{t+1}}\left\{\sum_{t+2}^{T} c_s^{\gamma} \mid y^t\right\}. \tag{A8}$$

Substitution of (A7) and (A8) into (A5) proves the theorem.

References

1. STRIEBEL, C., Sufficient statistics in the optimum control of stochastic systems, *Journal of Mathematical Analysis and Applications*, Vol. 12, pp. 576–592, 1965.
2. WITSENHAUSEN, H. S., A standard form for sequential stochastic control, *Mathematical Systems Theory*, Vol. 7, pp. 5–11, 1973.
3. FELDBAUM, A. A., *Optimal Control Systems*, Academic Press, New York, 1965.
4. BAR-SHALOM, Y., and TSE, E., Dual effect, certainty equivalence, and separation in stochastic control, *IEEE Transactions on Automatic Control*, Vol. AC-19, pp, 494–500, 1974.
5. KENDRICK, D., Adaptive control of macroeconomic models with measurement error, in *Optimal Control of Econometric Models*. Edited by S. Holly, B. Rustem and M. B. Zarrop, Macmillan, New York (forthcoming).
6. WITTENMARK, B., Stochastic adaptive control methods: A survey, *International Journal of Control*, Vol. 21, pp. 705–730, 1975.
7. MANDL, P., Estimation and control in Markov chains, *Advances in Applied Probability*, Vol. 6, pp. 30–50, 1974.
8. ELLIOTT, R. J., The optimal control of a stochastic system, *SIAM Journal on Control and Optimization*, Vol. 15, pp. 756–778, 1977.
9. THOMPSON, G. L., Signaling strategies in n-person games, in *Contributions to the Theory of Games II*. Edited by H. W. Kuhn and A. W. Tucker, Princeton University Press, Princeton, New Jersey, 1953.
10. SANDELL, N. R., JR., and ATHANS, M., A finite-state, finite-memory minimum principle, *Journal of Optimization Theory and Applications*, Vol. 25, pp. 289–306, 1978.
11. HO, Y.-C., KASTNER, M. P., and WONG, E., Teams, signaling and information theory, *IEEE Transactions on Automatic Control*, Vol. AC-23, pp. 305–312, 1978.
12. RISHEL, R., Necessary and sufficient dynamic programming conditions for continuous-time stochastic optimal control, *SIAM Journal on Control and Optimization*, Vol. 8, pp. 559–571, 1970.
13. DAVIS, M. H. A., and VARAIYA, P., Dynamic programming conditions for partially observable stochastic systems, *SIAM Journal on Control and Optimization*, Vol. 11, pp. 226–261, 1973.
14. STRIEBEL, C., Martingale conditions for the optimal control of continuous-time stochastic systems, unpublished, 1974.
15. SAMUELSON, P. A., Proof that properly discounted present values of assets vibrate randomly, *Bell Journal of Economics and Management Science*, Vol. 4, pp. 369–374, 1973.
16. SAMUELSON, P. A., Lifetime portfolio selection by dynamic stochastic programming, *The Review of Economics and Statistics*, Vol. 51, pp. 239–246, 1969.
17. SAMUELSON, P. A., and MCKEAN, H. P., JR., Rational theory of warrant pricing, *Industrial Management Review*, Vol. 6, pp. 791–810, 1964.
18. BENES, V. E., Existence of optimal strategies based on specified information, for a class of stochastic decision problems, *SIAM Journal on Control*, Vol. 8, pp. 179–188, 1970.

Index

Affine structure, 209
Aggregate signal, 185
Allocation function, 233
Allocative process, 231
Arbitrage, 137, 175
Arrow-Debreu model, 15, 22
Asset pricing, 5
Asset-pricing function, 20

Bargaining process, 232
Bellman equation, 24
Bias of decentralized control, 238, 240
Budgetary process, 224
Bureaucratic behavior, 223

Cartel, 133
  cheating strategy, 144
  equilibrium, 134
  model, 135, 142
Chattering of solution, 79
Choke price, 136
Cobb-Douglas utility function, 42, 60
Common property
  exploitation, 168
  fishery resource, 117
Consumption dynamics, 59
Contraction mapping principle, 21, 209
Coordination of policies, 244, 245
Cournot-Nash oligopoly, 133

Decentralized controls, 179, 183, 237, 253
Differential game, 122
  pursuit-evasion, 73
Duopoly market, 216
Dynamic programming, 53, 109, 166

Elasticity, 135, 147
Envelope theorem, 112, 178
Equilibrium
  bionomic, 120
  consumer, 17
  distribution, 185
  hierarchical, 207, 210
  market, 133, 138
  noncooperative, 148
  point, 61, 93
Exploratory effort, 107

Felicity function, 55
Firm
  net worth of, 65
  theory of, 31
Fishery
  regulation, 127
  resource, 117

Fixed-point problem, 20

Game
  of brinksmanship, 233
  of economic or social behavior, 96
  stochastic, 91, 95
Gaps, inflationary and deflationary, 75, 81
Gateaux variation, 207
Generalized eigenvalue, 241
Growth model, 7
  neoclassical, 54

Hierarchical decisionmaking, 158, 205
Hierarchical equilibrium, 207, 210

Information set of agents, 228
Information structure, 179, 207, 210
  decentralized, 237, 260
Investment behavior, 62

Klein's Model I, 242

Large-scale systems, 179, 183
Leasing arrangements, 169
  risk-sharing, 177
Lobbying effort, 236
Locally-in-time optimal, 258
Lyapunov equation, 85, 209
Lyapunov function, 73, 85

Macroeconomic system
  decentralized, 237
  uncertain, 71
Marginal conditions, 67
Markov decision process, 95
Martingale (supermartingale), 257
Maximum economic yield, 120
Mean ergodic theorem, 14
Median voter result, 230
Minimum principle, 258, 260
Mismatched systems, 84

Nash, bargaining problem, 233
Nash competitive equilibrium, 121, 123, 129, 136, 144
Nash game, 95
Noncontrollable systems, 74

Oligopoly
  noncooperative, 146
  pricing, 133
  theory, 89
Optimal allocation, 10

Pareto optimum, 60, 123, 164
Person-by-person optimal, 261
Poisson distribution of mines, 107
Pontryagin maximum principle, 34, 52, 159
Preference structure, 91
Pricing dynamics, 90
Principle of optimality, 256
Production function, 63, 125, 191, 227
Profit maximization, 19, 31

Rational expectation, 14
Rational expectations equilibrium, 17
Rental market, 15
Resource
  allocation, 52, 89, 223
  extractive, 104
  exhaustible, 161
  fishery, 117
  nonreplenishable, 133
Royalties, 169

Security market line, 14
Shadow price, 112
Shadow rent, 111
Slutsky-type equation, 234
Solution
  funnel, 77, 84
  ultimate boundedness of, 76, 83
Stability, global uniform asymptotic, 78
Stabilization, 73, 237
Stackelberg solution, 158, 206
Stackelberg strategy, 212
Stationary policies, 92
Steady-state solution, 23, 60, 193
Stochastic difference equation, 11, 196
Stopped process, 162
Strategy, mixed, 232
Subsystems, 183
Sufficient static, 253
Supergame, 92

Targets, 73
  conflicting, 239
Tax price, 227
Team decisionmaking, 179, 255
Threat budgets, 231
Transversality condition, 52
  at infinity, 11, 18, 60, 164

Utility function, 11, 34, 109, 162, 229

Welfare functional, 226

Yugoslav firm, 44

MIX
Papier aus verantwortungsvollen Quellen
Paper from responsible sources
FSC® C105338

If you have any concerns about our products,
you can contact us on
ProductSafety@springernature.com

In case Publisher is established outside the EU,
the EU authorized representative is:
Springer Nature Customer Service Center GmbH
Europaplatz 3, 69115 Heidelberg, Germany

Printed by Libri Plureos GmbH
in Hamburg, Germany